Risk, Uncertainty and Decision-making in Property Development

JOIN US ON THE INTERNET VIA WWW, GOPHER, FTP OR EMAIL:

WWW: http://www.thomson.com
GOPHER: gopher.thomson.com
FTP: ftp.thomson.com
EMAIL: findit@kiosk.thomson.com

A service of

OTHER TITLES FROM E & FN SPON

Industrial Property Markets in Western Europe
Edited by B. Wood and R. Williams

Investment, Procurement and Performance in Construction
The First National RICS Research Conference
P. Venmore-Rowland, P. Brandon and T. Mole

Land and Property Development
New directions
Edited by R. Grover

Management, Quality and Economics in Building
P. Brandon and Z. Bezelga

Marketing the City
The role of flagship developments in urban regeneration
Hedley Smyth

Microcomputers in Property
A surveyor's guide to Lotus 1-2-3 and dBase IV
T.J. Dixon, O. Bevan and S. Hargitay

The Multilingual Dictionary of Real Estate Terms
A guide for the property professional in the Single European Market
Edited by I. van Breugel, B. Wood and R. Williams

National Taxation for Property Management and Valuation
A. MacLeary

Negotiating Development:
rationales and practices for development obligations and planning gain
P. Healey, M. Purdue and F. Ellis

Property Development: fourth edition
David Cadman and Rosalyn Topping

Project Management demystified:
second edition
Geoff Reiss

Property Investment and the Capital Markets
G.R. Brown

Property Investment Decisions
A quantitative approach
S. Hargitay and M. Yu

Property Management: second edition
D. Scarrett

Property Valuation
The five methods
D. Scarrett

Rebuilding the City
Property-led urban regeneration
Edited by P. Healey, D. Usher, S. Davoudi, S. Tavsanoglu and M. O'Toole

Risk Analysis in Project Management
J. Raftery

Urban Regeneration, Property Investment and Development
J. Berry, W. Deddis and W. McGreal

Effective Speaking
Communicating in speech
C. Turk

Effective Writing
Improving scientific, technical and business communication
2nd edition
C. Turk and J. Kirkham

Write in Style
A guide to good English
R. Palmer

Journal

Journal of Property Research
Edited by Bryan D. MacGregor (UK), Pat Hendershott (USA)

For more information on these and other titles please contact:
The Promotion Department, E. & FN Spon, 2–6 Boundary Row, London SE1 8HN.
Telephone 0171-522 9966.

Risk, Uncertainty and Decision-making in Property Development

Second Edition

Peter Byrne

Senior Lecturer
Department of Land Management
and Development
The University of Reading

Taylor & Francis
Taylor & Francis Group

LONDON AND NEW YORK

Taylor & Francis,
2 Park Square, Milton Park,
Abingdon, Oxon, OX14 4RN

First edition 1984

Second edition 1996

Transferred to Digital Printing 2005

Typeset in Times Roman 10/12 by Acorn Bookwork, Salisbury, Wiltshire

ISBN 0419 20030 4

A catalogue record for this book is available from the British Library

Library of Congress Catalog Card Number: 96–68683

To C. J. B.
who said

'Not more numbers. . . .!?'

Contents

Preface

With all the changes that have taken place in the property development industry since the first edition of this book appeared, it might have been thought that this second edition would be very different. This is not really the case! The principles have remained unchanged, and what was true in 1984 is still true now. Thus the first part of the book has been added to and amended only to clarify areas which were felt to be a bit limited or obscure in exposition.

The advent of spreadsheets, soon after the first edition appeared, revolutionized business computing. It also transformed our ability to do development appraisal calculations, and the latest generation of spreadsheet packages allows the analyst previously undreamt of levels of 'sophistication' in the approaches that are possible. Substantial changes have therefore been made to Chapter 5, and a new Chapter 6 has been added to reflect these developments.

The main consideration which has affected the changes made overall, however, has been the views of many readers of the first editon, who have uniformly suggested that the structure (and indeed most of the content) should be left alone! In spite of my enthusiasms, I have tried therefore to do that.

David Cadman contributed substantially to the first edition of this book. Although he was not directly involved in this new edition, his spirit, and at least some of his words remain, with my thanks.

Additional thanks go to colleagues in the Department at Reading, in particular to Roger Gibbard for his development appraisal example, used in Chapters 5 and 6. Thanks also to my long suffering editor, Madeleine Metcalfe at Spon, who despairs, quietly and patiently, of academics (trying to write books).

Preface to the first edition

Whenever a decision is made, whether it is to take an umbrella to work in case it rains, or to invest £10 million in a hotel development, that decision is, in a sense, nearly always the result of a compromise. This is because the decision-maker has to decide whether the information currently available is adequate to make it 'worth' taking the umbrella, or investing £10 million. (But not as alternatives!) The process of deciding could be totally intuitive, pragmatic and 'seat of the pants', it could be very rational, very scientific, very considered, or it could be a compromise between these two extremes. It depends on the possible outcome of the decision. If you fail to take the umbrella, and it rains, you probably only get wet. If you fail to make an adequate return on the £10 million invested, the consequences *could* be more serious. That might encourage a more considered approach at least to the latter decision. Even so, in neither case do we actually know what will happen. All that can be done in the time available is to attempt to discover, and minimize the effect of, those factors which may cause the decision to be wrong.

This book introduces ways in which this can be done, in the particular context of property development.

The book bridges the gap between some very theoretical and mathematical aspects of a growing subject area – the analysis of decisions given risk and uncertainty – and the very practical aspects of the day to day process of property development. Inevitably it may seem to some that important aspects of each of these areas are dealt with in a way which is less than complete. The intention has, however, been to encompass sufficient of the technicalities of each area to enable workable links to be made at an essentially practical level.

Uncertainty, risk and the process of property development

1

1.1 INTRODUCTION

John Kenneth Galbraith writing in the 1970s called the latter part of the twentieth century *The Age of Uncertainty* (Galbraith, 1977). His theme was to 'contrast the great certainties in economic thought in the last century with the great uncertainty with which problems are faced in our time'. Once this uncertainty is accepted as being, as it were, in the nature of things, it is easily seen that those who enter into ventures that depend upon future outcomes have to come to terms with it. One such group of people are those who control business ventures. Consequently it is not surprising to find that it is in this area of activity where attempts have been made to develop a theory of decision-making that copes with uncertainty. Decision analysis has now become a well established part of the curricula of business schools and the like, although it is fair to say that to date the basis of that theory has tended to support a mostly rational/quantitative methodology. These ideas are only beginning to gain a more general acceptance, particularly with the development of techniques such as scenario analysis and more recently and still controversially, fuzzy analysis.

At first sight it might be supposed that this body of theory and practice would have been taken up enthusiastically by those concerned with the development of land and buildings. After all, uncertainty lies at the root of the process of property development which is essentially concerned with the manufacture of a product in anticipation of an unknown future demand. Indeed, if it were not for the constraint upon supply imposed by the system of town planning, it would rank as one of the most speculative of activities, involving as it does the investment of relatively large amounts of capital into a product that is fixed both in time and space. Yet the property development industry has largely ignored the methods of formal decision analysis adopted extensively in other industries (see for example Marshall and Kennedy, 1992).

Why is this so? It is, no doubt, due in part to the entrepreneurial nature of the business and to the fact that most development companies are, in terms of numbers employed, comparatively small. The formulation and implementation of policy does not therefore require much of a formal structure. Even the largest property companies or the property departments of financial institutions, such as insurance companies and pension funds, are small in management terms when compared with most industrial companies.

There are, however, reasons to suggest that a more formal approach to decision-making is important.

First, property markets internationally have become essentially institutionalized. The financial institutions in general and the pension funds and insurance companies in particular have come to have an overwhelming influence upon the kind of development that is undertaken and the way in which that development is carried out. For a description of the growth of financial institutions in property in the post-war period see Cadman and Catalano (1983), Cadman (1984) and Plender (1982). The informal entrepreneurial spirit of the small property company is much less suited to the perceptions of these institutions in which there is a much greater need to explain and justify particular courses of action.

Second, the collapses of property markets in 1974/5, and again at the end of the 1980s exposed a (continuing) paucity of analysis upon which investment had been made and some schemes had been carried forward, and showed that some lessons had been hardly learnt by the industry. The collapse of large and apparently well founded development companies in the latter period show at least in part the need to have a strong understanding of the factors, however diffuse, that affect the development market (see Ross Goobey, 1992).

Third, the developments in computing hardware and software have made commonplace forms of analysis that were previously both too time-consuming and cumbersome. Fourth, the growth of a body of structured and systematic data is providing decision-makers with a much improved framework of market information. Finally, as more and more people become enmeshed in investment, particularly through pension schemes, the direction of investment is being subjected to much greater internal scrutiny and fund managers increasingly have to explain and justify their own performance and have forced the same levels of scrutiny on those presenting investment opportunities. However much faith is placed in 'the seat of the pants', it is difficult to explain it to the outside world.

1.2 THE PROCESS OF PROPERTY DEVELOPMENT

Even when related specifically to property, 'the development process' means different things to different people. To some it is simply the construction of buildings, a physical process of production. To others it is essentially a part of a social and political process, involving the distribution and control of resources. We do not seek to deny either of these interpretations and indeed accept that there are others that, while valid in themselves, are not appropriate to our purpose. In answering the question 'What is development'?, the Pilcher Report (HMSO Report, 1975) gave the following reply:

Development comprises the following tasks:

(i) The perception and estimation of demand for new buildings of different types;

(ii) The identification and securing of sites on which buildings might be constructed to meet that demand;

(iii) The design of accommodation to meet the demand on the sites identified;

(iv) The arrangement of short- and long-term finance to fund site acquisition and construction;

(v) The management of design and construction; and

(vi) The letting and management of the completed buildings.

This definition is close to the one that we need for our study. Our own definition is as follows:

The process by which development agencies, together or on their own, seek to secure their social and economic objectives by the improvement of land and the construction or refurbishment of buildings for occupation by themselves or others.

Acquisition, production and disposal

Although there is no model of the development process that can be applied universally, for the purpose of our study, which is principally the investigation of uncertainty and risk, we can divide the process into three parts as follows:

1. Acquisition
2. Production
3. Disposal

The first part of the process comprises the acquisition of the land upon which the development is to take place and the acquisition of the appropriate planning permission. The second part comprises the construction of the building or buildings and the third part comprises their disposal both for occupation and investment. As the process takes place, the developer's knowledge of the likely outcome increases but, at the same time, the room for manoeuvre decreases. Thus, while at the start of the process developers have maximum uncertainty and manoeuvrability, at the end they know all but can do nothing to change their product, which has been manufactured on an essentially once and for all basis. The process is especially susceptible to risk and uncertainty because, once started, it is relatively fixed in time and place and because it aims at a very narrow consumer market.

During the first part of the process, the period of acquisition, the uncertainties are of three main types: the physical characteristics of the land, the characteristics of tenure, including restrictive covenants and easements both in favour of and against the land, and the nature and extent of use that the local planning authority will permit. Most developers will attempt to identify and determine all these factors before committing themselves to the purchase of the land. Land acquisition cost is often the first major commitment of capital and as it comes at the start of the development process it is then like a snowball, accumulating interest throughout the remainder of the development.

During the second part of the process, the period of production, the main element of uncertainty is the cost of construction. This factor, which represents the second major capital commitment, is substantially determined at the start of the building contract with the builder. However, in nearly all cases there will be some element of fluctuation allowed for in the contract and, in any event, the phasing of the construction and the length of the building period can never be fully determined at the outset.

The third part of the process, the period of disposal, can be seen as comprising both the disposal of the building to one or more occupiers and its disposal as an investment. Where the occupier is also the owner, as in the case of the residential owner-occupier or where a commercial or industrial building is sold to a company for its own use, the two forms of disposal are fused together. In most commercial and industrial developments, however, the buildings are let to one or more occupiers on full repairing and insuring leases and the consequential stream of net income can then be sold as a property investment. The uncertain factors that lie within the disposal phase are, therefore, rent and investment yield or, in the case of residential property, capital price. Because it is physically fixed to a precise location, and because it has to be manufactured well in advance, the eventual product of property development is trapped within a particular social and economic framework which is largely beyond the developer's control. No developer of a speculative development project can be sure of the market conditions that will prevail as and when the development is completed.

Some of these uncertainties can be contained. As we have already noted, the purchase of the land can be delayed until planning permission has been received; building contracts can be agreed on a fixed price; and buildings can be pre-let or pre-sold. In such circumstances, developers typically have to trade off increased certainty against a lower level of profit. Land which has the benefit of an appropriate planning permission costs more than land for which it is only hoped that planning permission will be forthcoming; builders quote higher prices for fixed-price building contracts; pre-letting may mean agreeing a lower rent even in a rising market, or no rent at all in a falling or dead market; and investors expect to achieve a higher yield if they commit themselves to the purchase of an investment before it is completed and let. Every developer has to calculate the advantage to be gained from increased certainty and balance it against greater risk but higher potential gain, and the way in which this is done will be a function of the developer's attitude to risk.

Time

There are some kinds of uncertainty that are, to an extent, within the developer's control. They can be identified, measured and contained. Others are more difficult to control and to a large extent have to be accepted as part of the development risk. One such element of uncertainty is that of time. Property development is a dynamic process and time runs through it as a constant source of uncertainty. Property development is a time-consuming business. Few projects can be put together and carried out in less than a year and many take several years to complete. This being so, it is difficult, at the outset, to make accurate statements as to the actual flow of expenditure and income throughout the development period. They can only ever be estimates. Furthermore, because of the inevitable time lag between the conception of the project and its completion, property development is especially vulnerable to broadly based, and local, social, economic and financial changes. Changes in consumer preference, the rise and fall of economic cycles, or changes in interest rates are difficult to foretell over even the medium term and once a development has been started it is difficult to change. It can often happen, therefore, that the new supply of accommodation comes on stream at a time when the upsurge in demand that stimulated development has begun to decline, giving rise to a cyclical pattern of over- and under-supply which appears to be a regular feature of property markets (see for example Barras, 1979, 1994, Key *et al.* 1994).

Nevertheless, by identifying those elements of uncertainty that are to some extent within their control, and by recognizing the reality of uncontrollable uncertainty, developers can put themselves in a better position to make informed and measured decisions about their projects.

Attitudes to risk

The main 'developers' are property companies, financial institutions, public sector agencies and occupiers. Some of these developers are more inclined than others to accept uncertainty, or 'take a risk'. Typically, property development companies are thought to be the greatest risk-takers but the extent to which they feel comfortable with risk varies and most of them would actually deny that they take any risks without careful assessment. In the event, the constraints of finance mean that all serious projects have to be properly appraised and presented if they are to secure short-term development finance or long-term funding. It is intuitively recognized that variations in particular variable factors such as rent or investment yield affect the outcome of the development more than others. To this extent, an attempt is being made to assess the 'riskiness' of the project. As we shall see, however, present practices tend to depend too much upon the uncritical acceptance of apparently well informed 'best estimates' of income and expenditure without an adequate exploration or assessment of the range of possible and probable outcomes. Attempts will now often be made to test the development appraisal for variations in the values of the estimated variable factors such as rent and building costs, but this is seldom carried out with any degree of rigour or formal methodology. Any formal attempt to measure the sensitivity of any appraisal to the uncertainties involved remains a somewhat rare event.

The growth in substance and importance of the financial institutions, in particular the insurance companies and pension funds, has encouraged a greater sophistication in property development appraisal as these institutions have a hierarchical form of decision-making with a high degree of accountability. By their own definition, they are very much in the role of trustees: see, for example, the evidence given to the Wilson Committee by the financial institutions (HMSO Report, 1980). Their attitude to risk changed as they became more and more directly involved in property development. In the period immediately after the Second World War they limited themselves to the mortgage market but throughout the last thirty or so years they have become first investors and then developers and investors once again. The experience of the property collapse of 1974/5 made it clear to them that in the last resort they had to bear the burden of risk rather than the property companies to whom they had committed themselves. In the collapse of the late 1980s and early 1990s, the banks learned the same lessons. Perhaps because they had not had such a direct involvement in development activities they found themselves bearing the consequences of developer's risk-taking activities, without themselves ever having really considered what those risks might actually be. In many cases it would now be difficult to distinguish between the attitude to risk of the major property development companies and that of the major pension funds, insurance companies or banks, and their fortunes are in many ways intimately related.

Attitudes to risk in the public sector have been and are very mixed. In some cases local authorities have preferred to acquire land and then grant development leases to private sector developers in an attempt to limit their own risk. On the other hand, however, some public sector development agencies such as the new town development corporations have carried through major projects of property development entirely on their own. During the 1980s, however, increasingly severe financial constraints generally caused a withdrawal of local authority interest in large-scale development.

There is likely to be as great a range of risk aversion within these different categories of developer as there is between the groups themselves. As we shall see, these variations in willingness to take risks are an important element in the theory of decision-making.

Information

Well informed decisions and decisions that explicitly confront the problem of measuring uncertainty, depend substantially upon an adequate information base. The property industry was slow to develop such systematic information systems. Until quite recently there was little serious attempt to provide a quantitative assessment of the main variables of rent, cost and yields and there is still an argument as to the appropriateness of current measures. It was traditional to depend upon the 'experience' of the particular professional adviser such as the estate agent, valuer, architect or quantity surveyor and those advisers were inclined to revel in the mystique of their own profession or even deliberately to promote themselves as being able almost intuitively to 'have a feel for' market conditions.

Over recent years, however, an increasing number of data sets have been produced measuring the performance of rents, building costs, investment yields and house prices and there is little doubt that these services will continue to expand and become more sophisticated, reliable and authoritative. As they do, better attempts will be made to do more than simply record past events. Gradually, the existing indices that measure trends are being projected forward to give forecasts of possible future values. If the underlying research is done carefully and if the information is presented in a readily digestible form, then these new services can provide developers with valuable information which will enable them, and their advisers, to make better assessments of uncertainty and risk.

The development of information systems and the analysis and presentation of information have been greatly assisted by the availability of powerful computers. As we shall see, these machines are also of great assistance in formal decision-making and in complex development appraisal. The development of computer software such as spreadsheets makes it possible to carry out sophisticated development appraisal with little apparent effort. In addition, a number of development appraisal packages of varying scope are also to be found in the specialist real estate software market (see Chapter 4).

Uncertainty and risk

So far, we have been using the terms 'uncertainty' and 'risk' in a rather colloquial form. If, however, we wish to take our study forward on a more rigorous basis we must define these terms somewhat more precisely.

For our purpose, uncertainty is taken to be anything that is not known about the outcome of a venture at the time when the decision is made. In contrast, risk is taken to be the measurement of a loss, identified as a possible outcome of the decision. 'Loss' need not be measured in purely monetary terms and may be as much perceived as actual. As we shall see, the adoption of these interpretations of uncertainty and risk has the advantage of making a distinction between the uncertain variables in the decision model and the riskiness of the project as a whole as perceived by the decision-maker.

1.3 DEVELOPMENT APPRAISAL

Having introduced uncertainty and risk, and having seen how they can impinge on the property development process, we can now begin to look more closely at the theory of decision-making and risk analysis. Before doing so, however, it may be helpful to look at an example of a typical 'traditional' development appraisal. This form of development appraisal is not particularly complex. Indeed, such a traditional approach might now be considered as too simple, and lacking in detail, but it serves well to identify the real uncertainties that lie within the system and are unexplored.

Consider the following example. A developer is offered five acres (2 ha) of land with detailed planning permission for 100 000 ft^2 (9290 m^2) of warehousing. The site comprises a level and rectangular area of land at the periphery of a provincial market town. It has access to the main road network and all main services are available. The asking price for the freehold interest in the land is £500 000.

In order to keep our example as simple as possible, let us recognize the potentially uncertain factors which we have obscured or eliminated by our assumptions. First, the interest that is being offered for sale is a freehold interest. The developer can study the title deeds to discover any onerous or beneficial rights of way, easements or restrictive covenants. Second, the land has planning permission in detail for a particular type and quantity of use. The developer can study the permission to discover any conditions to which it may be subject. Third, the land is level and rectangular. We further assume that there are no adverse soil conditions, although in practice this might require a soil survey. Fourth, the land has adequate road access and all main services are available.

Given these important assumptions, the calculation in Table 1.1 represents a development appraisal for such a modest sized commercial or industrial scheme. These calculations can be done manually (on the back of an envelope), but would normally be made, as here, using spreadsheet software.

While there is no standard form of development appraisal, this is typical of the traditional 'deterministic' method still often used in practice. It is built around the single figure, best estimates or 'point estimates' of the various advisers within the development team. Each figure within the appraisal, for example the building cost or rental income, is presented as a fixed and certain figure. However, the reality is that these fixed single figure factors or point estimates disguise the true uncertainty that lies behind the adviser's estimate.

We are able to identify within this appraisal thirteen factors which are truly variable. These are the 'variable factors' (a)–(m). Having isolated the thirteen variable factors upon which this appraisal is substantially dependent, let us look at each of them separately.

The variables

(a) Land

In our example we have assumed that the land cost is determined at the 'asking price'. In practice, if our appraisal has not produced a satisfactory developer's yield or profit, either the scheme would have had to be abandoned or some means found to reduce development costs. Because the land cost is a residual cost, it is susceptible to negotiation. Methods of residual valuation exist for determining land value, and it is common for negotiations to take place between the landowner and the developer until a bargain is struck at a price which can be satisfactorily inserted into the kind of development appraisal that we have shown. A residual calculation will form the basis of the examples in Chapters 5 and 6.

(b) Ancillary costs

The ancillary costs of land acquisition are comparatively easy to assess, being made up of agent's fees, legal costs and stamp duty. Each of

Table 1.1 Development appraisal

Land cost

1. Asking price for land	£500 000 (a)	
2. Cost of acquisition i.e. agent's and legal fees plus stamp duty @ 3.50%	£17 500 (b) £517 500	
3. Interest on land held from date of purchase to disposal say 15 months (c) at 17.00% (d)	£112 213	
Total land cost		**£629 713**

Construction cost

1. Building cost 100 000 ft^2 (e) at £16 (f) per ft^2 (including cost of internal roads, services and landscaping)	£1 600 000	
2. Professional fees at 14.00%	£224 000 (g) £1 824 000	
3. Interest for 12/2 months plus 3 months (h) at 17.00% (i)	£227 938	
Total construction cost		**£2 051 938**

Letting cost

1. Agent's fees at 15.00% of rental income	£37 500 (j)	
2. Advertising cost	£5000 (k)	£42 500
Total development cost (TDC)		**£2 724 151**

Rental income

100 000 ft^2 @ £2.5 per ft^2 per annum (l)		**£250 000**

Developer's yield

Rental income/TDC × 100		**9.18%**

Capital value

Rental income valued on the basis of an investor's net yield of 7.00% (m)		**£3 571 429**

Developer's profit

Capital value minus total development costs expressed as a percentage of the total development cost		**31.10%**

---

these is either the subject of a fixed scale or can be determined fairly accurately prior to purchase. In any event they comprise a very small part of total development cost. In our example, they comprise just over 0.5%.

(c) Land holding period

The period for which the land will be held and therefore the period for which it has to be financed, is made up of three parts:

1. The period between the purchase of the land and the start of construction. In the case of our example this may be quite easy to determine as planning permission is assumed to exist. In many cases this period will be very short, as developers will try to arrange the purchases so that building can begin at once. Where no planning permission exists this may mean entering into a contract to purchase subject to planning permission being granted. It may, however, be necessary to compare the advantage of buying land without planning permission at a lower price with the advantage of purchasing when permission is granted at a higher price
2. The period of construction to which we refer under (h) below.
3. The period for letting or selling the completed product. In our example we have assumed that no rental income is receivable before the completion of construction, and that in fact it takes a further three months to secure a tenant or tenants and arrange for leases to be granted. Where the scheme comprises a number of users it will normally be found that some are let before others, and the three month figure in our example is therefore something of an average Some schemes are organized so that parts are let or sold before the entire development has been completed, thus generating an income flow to offset the flow of expenditure. Phasing of this type may be very important in making some projects appear viable.

(d) Short-term rate of interest

Having established an estimated period for holding the land, it is necessary to apply an estimated rate of interest on short-term borrowing, to arrive at the cost of holding the land. The developer may have an established source of funds or he may have to arrange specific short-term funds for the particular scheme. The estimated rate of interest will reflect current rates and possible future trends.

(e) Building area

In our example we have assumed a gross building area of 100 000 ft^2 (9290 m^2). Where detailed planning consent exists this may be a matter of fact. Otherwise it has to be based upon discussion with the planning authority and a consideration of what is physically possible.

(f) Building cost

The estimate of the cost of construction is of great importance. Building cost, which may vary widely with inflation and the competition for work within the industry, constitutes a major component of development cost.

(g) Professional fees

An addition to the basic cost of construction is the cost of professional fees. Given the estimated building cost they can be calculated fairly accurately as they are normally related to established scales of charges, for example those of the RIBA and RICS. Our figure of 14% is made up as follows:

Architect	6%
Quantity surveyor	4%
Engineer	2%
Total	12%
Plus VAT at 17.5%	2.1%
	14.1%

In our example this item amounts to 8.2% of the total development cost.

(h) Building period

The length of the building period is important because it affects the cost of holding the land, and the cost of short-term building finance. The normal convention is to assume that the building cost is spread evenly over the building period and so in this example we have halved the period of twelve months. The developer would need to alter this calculation if it were known that this conventional spread was not appropriate to the particular case.

The cost of construction has to be financed during the building period and for the period it takes to sell or let the completed product. In our example we have therefore added to the building period of twelve months the letting period of three months (see (c) above).

(i) Short-term interest rate

We have already commented upon the rate of short-term rates of interest in (d) above and make no additional comment here.

(j) Letting fees

The amount of the agent's letting fee (or selling fee if this was, for example, a housing development) is fairly easy to assess as it is normally charged in relation to some established scale of charges. In our example we have assumed two agents, perhaps a national and a local agent, charging three-quarters of the scale recommended by RICS. The fee depends upon the estimate of rent and we refer to this in (l) below.

(k) Advertising costs

The cost of advertising includes the cost of advertising the property and of preparing and distributing particulars. There may be a quite simple 'single sheet' or in the form of a more elaborate and costly brochure. This item of cost is fairly easy to estimate and is normally a small percentage of total development cost. In our example it represents 0.18%.

These first eleven variables (a)–(k) all relate to development cost. In our example the basic building cost accounts for some 58.7% of total development cost and when added to the basic land cost and cost of short-term borrowing comprises almost 90% of total development cost.
　　The remaining two variables relate to the value of the completed development.

(l) Rental income

The rental income is an estimate of the rent that will be achieved when the development is completed, i.e. in our example, some fifteen months hence. The convention is that appraisals are based upon 'today's' costs and rental levels without allowing for inflationary (real) increases over the period of the development project. Indeed, the conservative convention is to allow for some inflationary effect on building costs but to keep rental values to current levels. The rental income directly determines the developer's yield and indirectly the developer's profit as it is a function of capital value.

(m) Investment yield

When the estimate of rental income has been established the capital value of the completed product is arrived at by capitalizing that income at the appropriate investor's yield, making allowance for the cost of disposal. The prime property investment market is compact and the flow of market intelligence comparatively 'efficient'. Yields do, however, vary over time and unless the investment is to be pre-sold (i.e. the terms for the sell-out are agreed with an investor prior to the start of the development) the developer has to form a view of the yields that will prevail at the end of the project, in our example some fifteen months away.

At this point we can summarize the variable factors in this example. They are as follows:

(a) Land cost
(b) Ancillary costs of acquisition
(c) Pre-building contract, building and letting period
(d) Short-term interest rate
(e) Area of building
(f) Building cost
(g) Professional fees relating to construction
(h) Building and letting period
(i) Short-term interest rate
(j) Agent's fees
(k) Advertising costs
(l) Rental income
(m) Investor's yield

Although the variable factors within a particular development appraisal will differ, this list is typical of a great many commercial and industrial development appraisals.

In our example, (a) and (e) are taken as 'known'. As we have seen, the following factors are normally so straightforward as not to represent any real uncertainty (together they amount to 10.4% of total development cost):

(b) Ancillary cost of acquisition
(g) Professional fees relating to construction
(j) Agent's letting (or selling) fees
(k) Advertising costs

That leaves the following principal variable factors:

(c) and (h) Pre-building contract, building and letting period
(d) and (i) Short-term interest rate
(f)   Building cost
(l)   Rental income
(m)   Investor's yield

Factors (c) and (h) and (d) and (i) can be combined as a single variable (the cost of short-term borrowing on both land and building costs), thus producing four main variable factors:

1. Short-term borrowing cost
2. Building costs
3. Rental income
4. Investor's yield

It is the interplay of these factors that substantially affects the success or failure of most development projects. It should be recognized that estimates of the time taken at various parts of the process are crucial to each of these four factors.

Sources of data

Although developers must in the end form their own judgements about the variable factors, they rely, in part, upon 'in house' or external professional advice. The following are the typical sources of advice for each variable factor:

Factor	*Source*
(a) Land cost	Either asking price or valuer/estate agent
(b) Ancillary cost of acquisition	Valuer/estate agent and solicitor
(c) Time periods	
(i) Pre-contract and building periods	Architect and quantity surveyor
(ii) Letting period	Valuer/estate agent
(d) Short-term interest rates	Developer
(e) Area of building	Architect
(f) Building cost	Architect, quantity surveyor and engineer
(g) Professional fees	Architect, quantity surveyor and engineer
(h) Time periods	
(i) Building period	Architect and quantity surveyor
(ii) Letting period	Valuer/estate agent
(i) Short-term interest rate	Developer
(j) Agent's fees	Valuer/estate agent
(k) Advertising costs	Valuer/estate agent
(l) Rental income	Valuer/estate agent
(m) Investor's yield	Valuer/estate agent

It should perhaps be noted that the developers themselves may play a substantial part in determining these estimates, particularly when they specialize in a particular type of development or locality and have built up a sound basis of experience. Where of course the builder and developer are one, the estimate of building costs will be largely 'in house'. Similarly, where the investor is also the developer there will be no need to estimate the investor's yield. Finally, where the occupier is the developer neither rental income nor investment yield needs to be estimated, except to the extent that the occupier needs to establish the market value of the completed building.

Sensitivity and uncertainty

The main variable factors that are most influential in development appraisal have now been isolated. It is only a short step to recognize that they are the most sensitive to change. A 10% adverse change in any one of the four main variables in our example is much more harmful than a similar adverse change in the remaining nine.

The recognition of sensitivity implies an acknowledgement of uncertainty. Development appraisals are prepared upon the basis of estimates because actual figures are not available. Rents and prices, for example, are estimated because they will not be determined until the project is completed and let or sold. Traditionally, estimates are deterministic, that is to say the members of the development team separately or together pick their best estimates of what the uncertain factor will be. The architect or quantity surveyor may advise on building cost, the letting or selling agent on rent or price and so on. In arriving at their estimates advisers may consider a range of possible figures, but will select the one that appears to be most likely. When the estimate has been arrived at on a 'today's' figure basis, some allowance may be made for inflation, although it is more common for such an exercise to be applied to building costs than to rental or capital values.

It is common for such an evaluation to be revised during the development process to take account of new evidence and recognized changes in inflation. Typically there are at least three such evaluations. The first is made at the outset when the project is first considered, to see whether the project appears to make any sense at all. Some people refer to such calculations as 'back of an envelope' calculations. The second evaluation takes place following a thorough investigation of the project by the members of the development team and is based upon a design and drawings prepared by the architect, after establishing the nature and extent of the development that will be permitted. The third evaluation is likely to take place immediately before the development commences when a substantial part of the total cost has been determined, the land price has been agreed, short-term finance has been arranged and a tendered building price received. Indeed, at this stage the degree of uncertainty may have been further reduced by a pre-letting and/or pre-sale.

As already noted, it is a common feature of the development process that as it proceeds the degree of certainty increases while the freedom for change is reduced. In common with all other 'rational' investors, it is also the case that the more risk that developers have to bear the greater is the return that they seek. Thus, if a project is unusual in design, type or location and if it has to be produced without having first secured an occupier, developers seek a higher profit or reward than if they were carrying out a familiar pre-let and pre-funded project on the basis of a fixed price building contract.

Traditionally, therefore, development appraisals are based upon a deterministic or at best a dynamic/deterministic assessment of uncertain factors. Developers and their advisers are aware that their estimates are uncertain and that some factors are more sensitive to change than others but there is little evidence of an attempt to measure the reliability of such estimates and therefore to determine more precisely the true measure of risk that is being undertaken. In considering variable factors, assumptions are made in determining the 'best' value. However, the 'best' value may not be the 'only' value for that variable. Thus while a letting agent may advise that the best estimate of rental value is £215 per m^2 (£20 per ft$^{2)}$ it would have to be acknowledged that it was at least possible that some other rental might in the event be achieved. One method of measuring the reliability of such estimates is known as probability analysis and it is this concept that is discussed in later chapters.

Finally, attention must be drawn to one further aspect of the traditional development appraisal. It has been shown that it misrepresents the true level of uncertainty by presenting variable factors as though they were fixed. It also fails to reflect correctly the cost implications of the timing of the factors of development cost. The most obvious example of this is the calculation of interest on building cost. The widely adopted method, treating building cost as though it were spread evenly over the building period [building period /2 × interest rate], is more often than not a distortion of the truth. In reality one needs to study the construction programme to establish the true pattern of expenditure (Baum, 1978). Furthermore allowance needs to be made for the fact that the payment of some part of the building cost will be deferred by way of a retention to cover the defects liability period.

There are a number of other such examples within the traditional appraisal and we will look at some of them in Chapter 5. The main purpose of this chapter is not to point to the mathematical inaccuracy implicit in the traditional approach but rather to reveal its inability to provide decision-makers with a proper indication of the real uncertainties in a project, thereby preventing them from being able to measure adequately the risks that they are taking.

<table>
<tr><td>**2**</td><td># Decisions and decision-making</td></tr>
</table>

2.1 INTRODUCTION

Having described the property development process and shown that conventional forms of appraisal do not cope adequately with the assessment of uncertainty and risk, we can now turn to the well established methods of decision and risk analysis to see to what extent they can assist.

The formal approach

Anyone familiar with the carrying out of property development will know that it contains many complex problems or, put another way, that there are many complicated decisions that have to be made. Some of these have been illustrated in the example in Chapter 1. In one way or another, objectively or subjectively, developers or decision-makers have to sort out or analyse the information that they are given and then make the best decision that they can. Because these decisions have to be made very largely on the basis of estimates or expectations of the future, the developer has to contend with uncertainty. The methods of analysis that we describe rely upon the development of formal, structured, 'pictures' of the decision problem sometimes called 'models'. These formal and systematic structures then lead to the application of sets of fairly rigorous criteria, particularly necessary when attempting to assess or measure uncertainty and risk. The result is that the intuitive feel for a particular decision can be amended by more precise and objective statements about outcomes under conditions of uncertainty. On this basis, the methods offer distinct advantages. On the assumption that they are applied correctly, they produce the following results:

1. They force decisions to be made in a logical and consistent fashion, with as much quantitative and qualitative precision as is possible given the constraints of time and resources. On the whole, this means that a

much more extensive analysis of the particular problem can be carried out when required.

2. The formal approach improves the attitude of the decision-maker to the quality of his own decisions, particularly where those decisions are usually of the intuitive kind. This is because the methods force the decision-maker to be much more specific about the criteria on which a decision is to be based, and to be consistent in the application of those criteria to successive decisions. If the decision-maker cannot be consistent, then he or she is forced to change that set of criteria or to accept its own internal inconsistency and consequences.

3. Such an approach enables errors to be traced, even if this is only with hindsight, thereby improving similar decisions at a later time.

It should be recognized that although this book is concerned mainly with development appraisal, the methods and techniques apply equally to other assessments or opinions of value made under conditions of uncertainty, including, for example, property valuations of all kinds and the assessment of investment opportunities.

The methods that are described make problems amenable to computer processing. This is important since some of the analysis of greatest value is difficult, although not impossible, without such assistance. The availability of such a computing capability is assumed in the discussion that follows, although in many cases it is not essential and its absence does not inhibit an understanding of the basic principles upon which the methods are based.

Perhaps the most important reason for considering these more systematic methods of development appraisal is that they provide the best, and in some cases the only, method for the proper consideration of the 'one off' kind of problem familiar in the development industry, especially those which are dealt with under conditions of uncertainty. It is important to recognize that although development appraisals have many common features, each development has elements that are unique and which make it necessary for it to be considered as a separate entity. This makes it difficult to develop a standard model, directly applicable in every case, and has tended to limit the accessibility of the approach.

Although emphasis will be laid on the advantages of these different approaches, this does not mean that they are being presented uncritically. There are many occasions where some or all of these techniques are inappropriate, akin to sledgehammers cracking nuts. Equally, there is no reason to suppose, automatically, that a decision that is based on a full, orderly and objective analysis is always better than a decision made entirely on intuition. In a particular case, the results may be identical or it may be that the intuitive decision will be shown to have been better. It is contended, however, that for most people, most of the time, the methods that are described are of real assistance in making better decisions. These techniques assist, but do not replace, the quality of judgement of the decision-maker. They are tools, useful tools, no more and no less. All sorts of decisions have to be made and some are comparatively straight-forward and would never necessitate the use of these methods. Never-theless, the use of formal decision methods can, with experience, alter the way in which all problems are viewed and lead to a much more consistent approach to their analysis and solution.

Complexity

Decisions can be simple or complex and they can be complex for a variety of reasons. First, they may be affected by external factors, the environ-ment in which they have to be made. For example, a development project comprising a new shopping centre may be constrained by multiple social and economic factors. The decisions that have to be made during the course of that development cannot be isolated from the social and economic environment but must take account of it. Second, decisions may be made complex by the nature of the decision-maker, that is to say by the internal character and structure of the decision-making organization. The organization may not have clear cut aims but a range of objectives and attitudes some of which may be in conflict with each other, or the internal process of an organization may be complicated by the fact that it is extensively 'democratic' rather than 'autocratic'. Third, decisions them-selves may be inherently complex. They may involve the consideration of a large number of variable factors and may present a range of solutions that are not easily distinguished. Such decisions need to be analysed with particular care to ensure that all the various influences have been properly identified and assessed. Lastly, decisions may be made complex because it is difficult to obtain, order and measure the 'intelligence' upon which they are to be made. This may result from the nature of the intelligence itself or from the limitation of resources of time and/or money in the pre-decision period. Much of decision analysis deals with these very large and complex kinds of decision, but generally this is beyond the scope of this book (see Keeney and Raiffa, 1976).

The formal approach to decision-making makes it easier to identify and unravel the real complexities and distinguish them from those that arise from 'muddled thinking'.

Kinds of decision

Although problems and decisions appear to vary enormously, they have certain common characteristics. There are two kinds of decision: single-stage or terminal decisions and multi-stage or sequential decisions.

Although the former may be internally complicated, they are made on the basis of information currently held by the decision-maker and do not lead on to further decisions. The latter, however, involve a sequence of stages before the final outcome is reached. This being so, it is helpful to be able to structure the process of decision-making. Both kinds of decision should be capable of being reduced to an analytical structure, i.e. they should be capable of being presented as a 'model'.

Models

A model is intended to provide a structured, usually simplified, but not necessarily over-simplified, representation of a problem and of the process of decision-making required for its solution. The objective of modelling is to enable a problem to be studied, analysed and adjusted, in order to arrive at the best solution. The performance of the model may be opti-mized not only internally but also in terms of the objectives of the decision-maker. The merit of the solutions that any model can give depends upon the extent to which the model can be regarded as truly representing the problem structure.

Three kinds of model are used: iconic models, analog models and symbolic or mathematical models. An architect represents a building by using an iconic model, a physical representation of a real building simply rescaled. When the board game Monopoly is played, a model of the property market is being used. This is not an iconic model, since the physical representation is not realistic. It is instead an analogue model. Some features of the real world are retained, money looks like money for example, but others are abstracted. This process of abstraction makes it easier to manipulate the model. At the greatest level of abstraction, generality and usefulness, are the symbolic or mathematical models which use symbols and mathematical functions to represent the variable factors in decisions and the way in which they interact. Models of this kind are often very complex in their own right, but need not be. The formula for compound interest, i.e. $A = (1 + i)^n$, is in a sense a mathematical model. It represents what will happen to one unit of something, usually money, when it goes through the process of being compounded at the rate of $i\%$ over the period of time n. Whatever the complexity of the system that is being modelled, it is worth attempting to devise as simple a structure as possible for the model. Such a model should be 'efficient' in terms of the time, cost and effort taken to develop it, capable of being widely inter-pretable and, if necessary, easy to alter.

Any model of the development process should be capable of including expressions of the uncertainty and risk associated with the variable factors of a particular development. They are likely to be of the more, rather than the less, complex kind. They are, however, most unlikely to be as complex as some of the most extensive mathematical models used for modelling physical systems. For example, the models of atmospheric circulation used for weather forecasting have required extensive research and development and need considerable amounts of computing time, even on the most powerful modern computers. By comparison, models of the development process are easy to devise, do not need extensive calculation, and are well within the capacity of a standard personal computer and a spreadsheet.

2.2 DECISION RULES

To develop the idea that different decision-makers may view the same problem in different ways, we can make use of a set of 'rules' which have been devised to describe the kind of decision generally to be expected from certain kinds of decision-maker. These rules were originally devised to describe decisions made under a condition described as 'true' or 'strict' uncertainty. Before examining how these rules are applied, we have to put aside temporarily our (practical) definitions of risk and uncertainty given in Chapter 1, and comment upon what this 'true' uncertainty is.

Consider first Table 2.1 which represents an apparently simple decision problem. The table consists of those elements of the problem which have been recognized by the decision-maker.

Suppose that a developer is trying to decide among four alternative schemes A, B, C and D. They are all considered to be rather speculative. The anticipated return for each will depend upon particular conditions over which the developer has no control. Such uncontrolled conditions are known technically as 'states of nature'. There are two states of nature represented in this simple example:

1. A rising market
2. A falling market

Table 2.1 A decision problem

Scheme	Rising market % return	Falling market % return
A	20	−11
B	17	−2
C	25	−6
D	8	4

What these markets actually look like does not really matter, except that the developer is able to say 'the market is rising' or 'the market is falling'. The returns for each of the four schemes will be affected accordingly and are different for each scheme under each 'state of nature'. There must be more than one such state of nature because if there was only one 'state' a condition of certainty would apply. It is assumed that the decision-maker does not know which state of nature will occur, only what they are.

The Pay-off table

A table such as this is called a 'decision matrix'; a 'decision table' or a 'pay-off table'. This pay-off table represents a state of 'true' uncertainty. The decision-maker is unable to say anything about the actual state of nature. His uncertainty about it cannot be measured in any way. The pay-off table therefore represents the outcomes, or pay-offs, of a particular course of action under a set of prescribed conditions, the states of nature. The pay-offs are often expressed in monetary terms, or, as in this case, in percentages but they can be in any unit that is meaningful to the decision-maker. This is important in those cases where the pay-off cannot be expressed in money terms. In such cases, the pay-off may be measured in units of 'satisfaction' or 'utility' (see Chapter 3).

In order to determine which is the 'best' scheme on the basis of his present knowledge, the developer has drawn up a table of returns for each scheme reflecting the rising or falling market. However the pay-off is measured, the table can now be examined for the best or optimum solution given the decision rule which the decision-maker, in this case the developer, adopts.

The maximin approach

This approach assumes a very pessimistic view on the part of the decision-maker. If things can go wrong, they will! The desirable decision is one that seeks to minimize risk and takes safety first and foremost. Thus the preferred course of action is that which gives the least bad of all the worst possible outcomes.

In our example, the thing that can 'go wrong' is the fall in the market. Searching through the pay-off table, the developer finds that scheme D minimizes the maximum loss or maximizes the minimum return. Under the maximin rule, therefore, if an investment is to be made at all it must be scheme D.

In many cases, this will lead to a 'don't invest' solution. Maximin represents a very, very conservative view indeed and any developer who attempted to use maximin consistently would undertake very few projects.

The maximax approach

If maximin is over-pessimistic, maximax is super-optimistic. It is the approach that appeals to those who positively like taking risks. The preferred course of action is that which gives the most favourable of all the best outcomes.

Adopting this approach in our example, the developer would choose to ignore the possibility that the market might fall. Assuming a rising market, he would search through the table for the schemes which maximized the best possible return. This is seen to be scheme C, returning 25%.

A rational approach?

In both of the previous cases, the developer has decided that one, and only one, state of nature can occur. In the case of maximin it is a falling market and in the case of maximax a rising market. In each case, the other state of nature is viewed as impossible. The decision-maker is assumed to be either pessimistic or extremely optimistic. These two approaches represent the end points of the risk spectrum, maximin – no risks taken, maximax – any risk taken. We do not, however, know anything about the states of nature, we do not know whether the market will rise, or fall. In reality, however, no developer expects to risk everything all the time nor does he always want to avoid risk altogether. There is no reason why any decision-maker should be completely optimistic or pessimistic. Indeed, most decision-makers fall between these two extremes. In the condition of true uncertainty, a rational approach might seem to be to declare that the states of nature are each equally possible and to consider their weighted outcomes to get a feel for the overall consequence of each action.

Since there are two states of nature here, treating them equally would assign a weight of a half to each. A weighted average can then be constructed by adding together the percentage returns multiplied by the equal weights, as follows:

(rising market pay-off × 1/2) + (falling market pay-off × 1/2)

The results of this are seen in Table 2.2.

Taking this approach, Scheme C is seen as performing 'best', but implicit in this performance averaging is the need to recognize that undertaking C will give a return of 25%, or –6%, but never the equally weighted average, 9.50.

Table 2.2 Pay-off table

Scheme	Rising market			Falling market			Weighted average
	% return	Weight		% return	Weight		return
A	20	1/2	+	−11	1/2	=	4.50
B	17	1/2	+	−2	1/2	=	7.50
C	25	1/2	+	−6	1/2	=	9.50
D	8	1/2	+	4	1/2	=	6.00

The Hurwicz approach

The next logical step, suggested by Hurwicz, involves a method that enables a decision-maker to state his or her degree of optimism on a scale from 0 to 1. Total pessimism is given a value (weight) equal to zero. This is equivalent to maximin. Total optimism is represented by a value (weight) of 1 and is equivalent to maximax. Since neither of these extremes is likely, the decision-maker is asked to consider the largest and the smallest pay-offs in the pay-off table and 'weight' them in accordance with his optimistic or pessimistic view. Applying an equal weight to each state of nature would be just one representation of optimism/pessimism.

Suppose that the developer in our example is twice as pessimistic as optimistic. He assigns a 'weight' of 2/3 to the minimum possible pay-off and a 'weight' of 1/3 to the maximum possible pay-off. A weighted average can then be constructed by adding together the percentage returns multiplied by the optimism and pessimism factor, as follows:

(rising market pay-off × weight) + (falling market pay-off × weight)

In our example this would appear as shown in Table 2.3. This points to a decision in favour of scheme D. The developer is rather pessimistic and is seeking safety. Scheme D gives the least damaging outcome in a falling market.

Table 2.3 Combining optimism and pessimism

Scheme	Rising market			Falling market			Weighted average
	% return	Weight		% return	Weight		return
A	20	1/3	+	−11	2/3	=	−0.67
B	17	1/3	+	−2	2/3	=	4.33
C	25	1/3	+	−6	2/3	=	4.33
D	8	1/3	+	4	2/3	=	5.33

Regret

A decision-maker may, in the event, regret her choice of action. She may wish that she had done something quite different. Savage (1954) called this experience 'regret' and suggested that a decision-maker should try to minimize his maximum regret. 'Regret' is defined as the difference between the pay-off from the best action and that of any other possible action, given the state of nature. To make a decision based upon this criterion, it is necessary to turn the pay-off table into a table of regret. This is done by subtracting each entry from the largest entry under the same state of nature.

If each entry under each state of nature in Table 2.1 is subtracted from the largest entry in that column the following revised table is as shown in Table 2.4. The largest entry in each column always has a zero regret as it is, of course, subtracted from itself. The scheme with the largest pay-off must always be the best course of action to choose. Choosing any other would mean that an amount of pay-off would be lost as a regret. This is an 'opportunity loss' and this term is sometimes used instead of 'regret'. If we are to follow this criterion, the maximum regret for each scheme has to be minimized. The maximum regret for each scheme in our example is given in Table 2.5. Minimizing these means selecting the lowest maximum regret value, which in this case is 8, and points to the adoption of scheme

Table 2.4 Table of regret

	Factor of regret	
Scheme	Rising market % return	Falling market % return
A	5	15
B	8	6
C	0	10
D	17	0

Table 2.5 Maximum regret

Scheme	Maximum regret factor
A	15
B	8
C	10
D	17

B as the 'best' scheme, that is, the scheme that best satisfies our developer's decision criterion.

It should be noted that some decision-makers, particularly the most cautious, might still prefer scheme D. Even though this has a 'regret' of 17 compared with scheme B's of 8, it is only an opportunity cost and might be preferred to the possibility of incurring a real loss of –2% in scheme B if the market falls.

True uncertainty implies that there is a fairly unrealistic 'world view' by the decision-maker. In effect it amounts to complete ignorance as to how the world, as measured by the states of nature is likely to perform. This is unrealistic and the condition of uncertainty is likely to be more partial.

These decision rules do highlight the attitude spectrum and show that different decision-makers' responses to the same situation can result in different decisions, each of which may be capable of rationalization. In that sense they are the bases for more developed methods of measuring the risk preferences of decision-makers. To develop the method it would be necessary to assign much more carefully measured values of optimism and pessimism. In addition, intermediate outcomes other than the best and worst may be important, but are not considered adopting this criterion. It may also be that all actions have the same best and worst consequences so that a particular course of action cannot be chosen.

When, or if, a situation is reached where the states of nature can be properly quantified, then the decision-maker has moved from a condition of uncertainty to one of 'risk'. In strict terms a decision is 'risky' if we are able to apply 'weights' to the states of nature which have been identified. Unfortunately hardly anyone using the word 'risk' is thinking in terms of properly quantifying the states of nature as they apply the word to their own circumstances. This is particularly evident in the case of property professionals of all kinds, where the word is used in a most indiscriminate way. It is for this reason that we have tried to define uncertainty and risk in ways which more clearly differentiate their meaning in relatively practical terms.

In this chapter, we have introduced the idea that it is possible to look at decision-making on a formal basis. The purpose of this is to make the process of decision-making more explicit and to enable decision-makers to analyse and order their decisions more effectively. We have introduced some of the methods and terminology of the formal approach. We have in particular introduced the concept of 'true' uncertainty and used it to show that different decision-makers can have different attitudes to the acceptance of risk. It is clear that these different attitudes result in variations in the selection of the 'best' course of action in given conditions. Different decision-makers looking at the same decision may prefer different solutions.

3	# Assessing risk and uncertainty

3.1 INTRODUCTION

Chapters 1 and 2 have introduced the idea of risk and have demonstrated that most ventures and decisions include elements of risk and uncertainty. It has been explained that uncertainty comes from the lack of predictability, or from the unsystematic behaviour of variable factors. The most common case is one in which variables exhibit unsystematic and unpredictable changes that occur at random and by chance. Any process that includes a random element of this kind is called a stochastic process. This can be contrasted with a process from which the same solution is always obtained if the basic conditions remain unchanged, which is called a deterministic process. The most important feature of a stochastic process is that its outcome is not predictable. It can differ from time to time and from place to place, even if the basic features remain unchanged. The process of property development has just these characteristics but is even more complex because its basic features do not remain constant. It can therefore be described as a complex stochastic process.

In order to be able to move from a development appraisal based upon conventional deterministic methods of the kind described in Chapter 1 to a more rigorous analysis that deals systematically with risk and uncertainty, it is necessary to become familiar with the terminology and methods conventionally used in such analyses. This chapter, therefore, deals with three aspects of analysis:

1. the measurement or assessment of probability;
2. the use of *utility* as an indicator of individual attitudes to risk;
3. sensitivity and simulation.

Perhaps the most important element in this is probability.

3.2 PROBABILITY

The language of probability

Probability

Probability is simply the way of measuring uncertainty. In any decision involving uncertainty, probability is used to describe the amount of uncertainty present. Often it is not actually called probability but is subsumed in a question such as 'Is it likely to rain tonight?' or 'What return is likely on this investment?' Most often it is expressed as a 'likelihood' and is automatically contained in non-specific answers ranging from 'very likely' to 'not sure' and 'don't know'. Most people therefore regularly deal with probability, but in this very unsystematic way. What is often recognized, however, is that situations where descriptions of this kind are used are 'risky', if only because the information that we have about the state of the world is imperfect.

On the other hand mathematicians and statisticians have developed very formal and abstract theories of probability. For our purpose, we need to adopt an approach that lies somewhere between the two extremes of the intuitive and the theoretical.

The random variable

In Chapter 1 we demonstrated that there are a number of variable factors within the development appraisal, elements such as rent and cost. The values of such factors vary between limits that may or may not be easy to fix. As we saw, in a conventional appraisal, the 'variable' is given a single value between those limits. Only through sensitivity analysis is any explicit attempt made to account for the fact that the variable may actually vary.

If the value of a variable is determined by events outside the control of the decision-maker, that is it is determined effectively by chance and probability, it is called a 'random variable'. This random nature of a variable can cause confusion. The randomness of a variable has nothing to do with the probabilities which are assigned to it. In general, probability is a numerical quantity, the value of which is determined by the outcome of a random experiment. The value that a variable takes is random because it results from events that cannot be entirely controlled, events such as the movement of the property market over time, or the effects of inflation on value. The random variable, in many different guises, is a fundamental part of any kind of analysis made under uncertainty.

The random experiment

Suppose that a developer is considering a project, the outcome of which is uncertain. As suggested in Chapter 2, in order to decide whether or not to pursue the project he needs to form a view of the range of possible outcomes (states of nature) and to attach a weight to each of them occurring. In more formal terms, the developer has to assign probabilities to the outcomes of a random experiment. The word 'experiment' is used to show that the outcome is not decided, and 'random' is used to show that each outcome is uncertain. The project may have a number of possible outcomes that *may* occur but one, and only one, will actually occur if they are mutually exclusive.

Consider the following classic random experiment. A coin is tossed. Excluding the possibility that the coin falls and stays edge on, there are two possible states of nature (outcomes) – a head or a tail. In this simple random experiment the list of possible outcomes is very restricted – a head or a tail. Most development projects are much more complicated 'experiments'. They have a larger number of possible outcomes. It is important to recognize, however, that the basic problem remains the same, i.e. the actual outcome is uncertain.

Measuring probability

Although we can now say that a random experiment or venture will have a range of uncertain outcomes, we have yet to say how uncertain the outcomes are. We have yet to measure the probability of each of the outcomes.

Some outcomes may be more likely than others. We may say, 'There is a good chance of achieving a rent of £18.00 per ft^2 (£1.67 per m^2) but a rent of £36.00 per ft^2 (£3.35 per m^2) is most unlikely'. Probability assessment gives a more precise meaning to 'a good chance' or 'most unlikely'. Conventionally, probability is measured on a scale from 0.0 to 1.0, these two values representing the limits of possibility. A probability of 0.0 indicates an outcome that cannot occur. An outcome that is certain to occur has a probability of 1.0.

How probabilities can be assessed

The relative frequency approach

The classical approach for measuring probability is the use of the random experiment. It is assumed that the experiment can be repeated. If it can, then the experiment is theoretically repeated an infinite number of times, or at least a very large number of times. The probability of a particular outcome occurring is then the number of times the outcome has occurred, as a proportion of the total number of experiments which have been carried out.

Continuing the coin-tossing example, suppose a coin is tossed 10 times, and a head shows eight times:

$$P(head) = \frac{\text{number of outcomes} = \text{head}}{\text{total number of outcomes}} = \frac{8}{10} = 0.8$$

where P(head) means the probability of a head.

Because the number of experiments here is small, a result such as this *is* possible. Since there are only two outcomes to this random experiment, fairly obviously, if the coin is fair we should expect tails to appear as often as heads. Heads should only appear half the time, and tails half the time. However, as a random experiment, just because a head has been tossed, it does not mean that a tail will be tossed next throw. It simply stands an equal chance of being thrown. In a short sequence of, say, ten throws, eight heads may therefore occur, and might be then followed by eight tails in the next ten throws. Twenty experiments will then have been made and if the probability is recalculated:

$$P(head) = \frac{10}{20} = 0.5$$

This answer although logically correct is still the result of a small number of experiments, and cannot be regarded as reliable. With small numbers of experiments the value of the probability calculated must be expected to vary, but will tend to stabilize as the number of experiments becomes larger and larger.

Clearly there will not be many occasions when anybody will be able to repeat an experiment, such as tossing a coin, a sufficiently large number of times to produce a stable probability. In some instances the number of occurrences can be observed in a large number of experiments. The probability of any woman giving birth to a boy can, for example, be determined by observing the number of boys born as a proportion of all children born in the population as a whole over say one year. In this case the experiment, giving birth to a boy, is repeated a sufficiently large number of times!

When observation is not possible or repeated experiment unreasonable, it is sometimes possible to determine the probability logically. Coincidentally, this has been done above in the coin-tossing example, and is most often done where the experiment can be shown to possess symmetrical (equally likely) outcomes. If the coin is not fair, biased perhaps towards heads, then the logical approach breaks down, and it is necessary to return to the long-run frequency method to determine the extent of the bias and the probabilities of each outcome. In such a case, the probability of tossing a head would be greater than 0.5.

Subjective assessment methods

There are many cases where probability cannot be assessed by the objective methods illustrated above because:

1. The experiment or venture cannot be repeated a sufficient number of times.
2. Insufficient information is available about the outcomes of past experiments.
3. The outcomes cannot be shown to be symmetrical.

Many business decisions are of this kind and property development decisions are no exception. Although probability may not be part of the developer's normal language, words such as 'likelihood', 'chance' and 'odds' often are. Using this language, the developer is in fact making *subjective* assessments of probability.

Any assessment of probability has to be based on the present state of knowledge. This knowledge can result from the accumulation of evidence of past events or it can be based upon that amalgamation of relatively abstract elements that is generally called 'experience'. Subjective assessment is dependent upon individual experience and will change over time as that experience changes. This also means that any two people can assign different probabilities to the same outcome just as, for example, two letting agents may have a different opinion of the chance of achieving a particular rent.

In any assessment of subjective probability the person making the assessment must be both honest and consistent. Although this may sound so obvious as not to be worth stating, experience suggests that a conscious effort and awareness are required to ensure that these two criteria are met.

There are many ways by which subjective probability can be measured: see Huber (1974) for a full discussion of these methods. Two fairly simple methods, known as the fixed and the variable fractile methods involve constructing a cumulative frequency curve from the answers given to a simple *questionnaire*. This has been extensively (and entertainingly) developed and discussed by Raiffa (1968).

The variable fractile method successively sub-divides the range of possible values into equally likely halves. In many respects this is to be preferred to the alternative method, the fixed fractile method, where the total range is successively divided into equally sized fractions but with different probabilities assigned to each. Although, like most things new, these methods are trying at first, they can be rapidly learned. After a short time the decision-maker should become familiar with them and be able to use them quite comfortably: see Wofford (1978). Furthermore, he should be able to see that decisions made in this way are 'good' in the sense that they can be seen to reflect faithfully his subjective beliefs.

The method, and its application, is discussed in more detail in Chapter 4, by reference to a specific example. The following section of this chapter describes the procedure in more general terms, showing the way in which a question and answer session might be conducted, and the way in which this is interpreted as a cumulative frequency curve.

A developer wishes to know the rental that could be expected from a new office letting. The letting agent is therefore asked to assess the possible levels of rent which might be achieved. The agent is questioned about his assessment as follows:

Questioner: What do you think are the highest and lowest possible rents that might occur?

Agent: Well, you can certainly expect £18.00 per ft^2, but I really can't see us getting £22.00. The market cannot exceed that level in the foreseeable future, and at the other end I can't really see it falling below £14.00.

Questioner: That gives us the possible range of rent. Now, what level of rent divides this range into two equally likely ranges? For example, do you think that the rent will be just as likely to go above £18.00 as below?

Agent: No, I'd rather take £17.00 as the 50:50 point, that is what you mean, is it not?

Questioner: Yes, very good. Now, consider demand levels below £17.00: if rent was to fall somewhere between £14.00 and £17.00 would you bet that it lay above or below £15.00?

Agent: Above, £16.25 seems more reasonable to me.

Questioner: That will do then as the 50:50 point. Let us now do the same thing for the upper portion.

Agent: I think that £18.25 is a sensible figure here. That means, doesn't it, that I think that £17.00 to £18.25 is as likely as £18.25 up to £22.00?

Questioner: Yes, that's it, the pattern is fitting together rather well so far, but now we need to make the scale rather finer. So, do you think that rent will be as likely to be between £16.25 and £18.25 as outside these limits? You have put 50% of the total range between these limits.

Agent: That seems about right: what would happen if I raised the upper limit to £19.00?

Questioner: Nothing, but why would you want to do this? At the moment you are betting heavily on a value close to £19.00. Are you unhappy about that now?

Agent: No, that is OK, I do think that values around £17.00 are most likely.

Questioner:	Right, suppose rent is above £18.25. What level splits this upper range into 50:50 segments?
Agent:	It must be £19.25.
Questioner:	Good, and above £19.25, on the same basis?
Agent:	Well, £20.50 is satisfactory.
Questioner:	At the other end of the scale, if rent is below £16.25?
Agent:	Down there, it could perhaps be £15.50. Yes, £15.50.
Questioner:	And then, below £15.50 can you split this segment in the same way?
Agent:	That is really the most difficult one that you've given me to think about. £14.00 itself is not really very likely at all even by comparison with £14.50. If I have to split it, let's simply say £14.75. Is that still satisfactory?
Questioner:	Splendid, this is now sufficient to enable us to construct a probability distribution describing your current beliefs about rents.

As a result of this question and answer process, the figures obtained can be formed into a table of cumulative probabilities. The table, as will be seen, does not have to be complete in the sense that the division of probabilities only needs to be sufficient to construct a usable probability distribution. Table 3.1 shows the breakdown into successively smaller halves (half of a half of a half, etc.).

Figure 3.1 shows the shape of the cumulative frequency in diagrammatic form. The irregular shape of the curve is a partial demonstration of the uncertainty inherent in the variable under consideration. In particular it helps the decision-maker to assess the reliability of a particular

Table 3.1 Subjective probability assessment of rent

Rents selected as 50:50 points	Cumulative probability
14.00	0.0000
14.75	0.0625
15.50	0.1250
16.25	0.2500
17.00	0.5000
17.50	0.6250
18.25	0.7500
19.25	0.8750
20.50	0.9375
22.00	1.0000

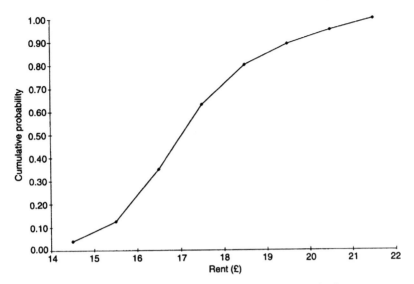

Figure 3.1 Cumulative frequency of subjectively derived rent

outcome, by telling him how likely the occurrence of that particular outcome appears to be. This will be discussed further in the next chapter.

These cumulative frequencies can now be reinterpreted as a frequency or probability distribution (a histogram).

In Table 3.2 the frequency distribution has been rearranged by comparison with Table 3.1, so that the classes are of equal width. This is done by taking those values directly from the graph (Figure 3.1). Inevitably this is an approximate exercise limited by drafting skill, but it is usually quite accurate.

Figure 3.2 is the relative frequency graph (histogram) for this example.

Table 3.2 Frequency distribution of rents: Equal class intervals

Rent interval (£)	Class mid-point	Cumulative probability	Class probability
14–14.99	14.5	0.040	0.040
15–15.99	15.5	0.125	0.085
16–16.99	16.5	0.350	0.225
17–17.99	17.5	0.630	0.280
18–18.99	18.5	0.800	0.170
19–19.99	19.5	0.890	0.090
20–20.99	20.5	0.950	0.060
21–21.99	21.5	1.000	0.050

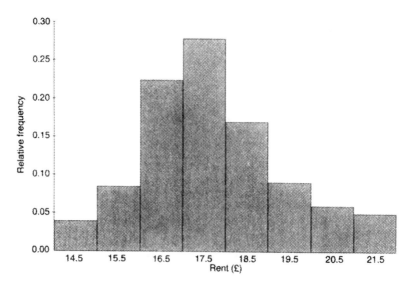

Figure 3.2 Histogram of subjectively derived rent (Figure 3.1)

Using this distribution, summary statistics, which describe the distribution numerically, can be calculated. Two measures in particular are important. These are the arithmetic mean, the average, and the standard deviation, the 'spread' of the distribution that is apparent from this graphical presentation. The most likely outcome, the mode, can also be determined.

The concept of expectation

One way of looking at probability is to say that it represents what will happen, in the long run, to the variable under consideration. It therefore provides a way of seeing how the variable might be expected to perform, on average. It should be noted how restricted this view actually is. By its very nature it is impossible to say how a random variable will perform unless a value of that variable can be assigned a probability of 1.0, i.e. its performance is certain.

Deriving an expected value

The measure usually used to describe the average performance of the distribution of a random variable is the arithmetic mean. If X is the random variable, and x_i ($i = 1, 2 \ldots n$) are the possible values which X can take, and if $P(x_i) = P(X = x_i)$ is the probability that X assumes the value x_i, then the mean is denoted by $E(X)$ and can be expressed by:

$$E(X) = \sum_{i=1}^{n} P(x_i)x_i$$

This equation represents the summation of all the possible values of X multiplied by their appropriate probabilities. The result, $E(X)$, is the mean of a probability distribution and is called the expected value. In cases where the data have been grouped the class mid-point (C) is used and the expected value is given by:

$$E(X) = \sum_{i=1}^{n} P_iC_i$$

where C_i is the ith class mid-point and P_i is the associated probability.

The expected value in the case of Table 3.2 is 17.21, which reflects the view of the expert that a rent of £18 was to be expected, but because of the distribution as a whole, is pulled down closer to the 50:50 point originally selected. There is evidence to suggest that decision-makers tend to produce subjective distributions which are not symmetrical, but fall away to the right-hand end of the distribution (they exhibit positive skewness), and in general terms this is what is observed here.

When decisions are being analysed, the expected value is extremely valuable in developing a criterion for making the best decision. In Chapter 2 the example shown in Table 3.3 was given. These 'weights' can now be described more correctly as probabilities and the weighted averages as expected values (or returns).

A criterion or rule, similar to those discussed in Chapter 2, may now be stated for making optimal decisions under conditions of uncertainty. This is Bayes'[1] decision rule (or Bayes' rule) which states that: 'In making a decision under uncertainty, the "best" decision is to choose that course of action which maximizes the expected return.' In the example above, therefore, choosing on this basis leads to action D, which has the highest expected value, pay-off or return. Often the expected pay-off will be expressed in monetary terms and in those cases it is usually known as the expected monetary value (EMV).

Table 3.3 Example of weights (probabilities)

Scheme	Rising market			Falling market			Weighted average
	% return	Weight		% return	Weight		return
A	20	0.33	+	−11	0.67		−0.67
B	17	0.33	+	−2	0.67		4.37
C	25	0.33	+	−6	0.67		4.33
D	8	0.33	+	4	0.67		5.33

It should always be remembered that an expected pay-off of this kind represents the supposed long-run *average* result of taking a particular decision and should therefore be seen only as a means of comparing such alternative actions as are available.

In our example, undertaking action D, the result will be either an 8% return, or a 4% return, but never the expected pay-off of 5.32, the probability weighted average of the two.

Using expectation to measure risk

In many cases, the use of the expected value will enable a sensible comparison to be made between outcomes. There will be occasions, though, where this is not possible. As an example, an institution is considering two investments A and B, and has already derived probability distributions for the possible returns from each investment (Table 3.4).

The expected returns are calculated for each investment:

$$E(A) = 11(0.2) + 12(0.6) + 13(0.2) = 12$$

$$E(B) = 10(0.1) + 11(0.2) + 12(0.4) + 13(0.2) + 14(0.1) = 12$$

How can a choice be made between investments when the expected returns are equal? A simple and often used solution is to consider the amount of variability, dispersion or scatter in each distribution, and to see how closely returns cluster around the expected value. Figures 3.3(a) and 3.3(b) compare the extent to which probability has been 'spread' across the possible returns from each investment.[2]

In this example, observation shows that returns from investment A are less spread than those from investment B, and as a result, actual returns from A are likely to be closer to the expected return than those of B. Investment A is therefore considered to be less risky, since the greater the spread of the distribution, the greater the riskiness of that variable.

This spread can be measured by the variance and associated standard

Table 3.4 Probability distributions

Return %	Probability of return	
	Investment A	Investment B
10	0.0	0.1
11	0.2	0.2
12	0.6	0.4
13	0.2	0.2
14	0.0	0.1
	1.0	1.0

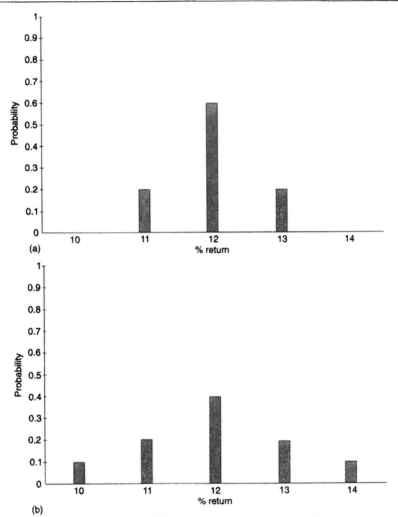

Figure 3.3 (a) Investment A; (b) investment B

deviation. The variance of the distribution, the square of the standard deviation, is calculated from:

$$\text{Var}(X) = \sum_{i=1}^{n} (x_i^2 P_i) - E(X)^2$$

Then:

$$\text{standard deviation (SD)} = \sqrt{(\text{Var}(X))}$$

Rewriting the variance formula given above to make calculation slightly easier:

$$\text{Var}(X) = \sum_{i=1}^{n} P_i(x_i - E(X))^2$$

The variances of A and B can be calculated:

$\text{Var}(A) = 0.2(11 - 12)^2 + 0.6(12 - 12)^2 + 0.2(13 - 12)^2 = 0.4$

$\text{Var}(B) = 0.1(10 - 12)^2 + 0.2(11 - 12)^2 + 0.4(12 - 12)^2 + 0.2(13 - 12)^2 + 0.1(14 - 12)^2 = 1.20$

standard deviation $(A) = 0.63$ standard deviation $(B) = 1.09$

Clearly B is much more variable than A, and given an equality of expected values, consideration of spread by calculation of variance will offer an alternative way of deciding between risky alternatives.[3]

Sometimes, rather than having identical expected values, distributions will have identical 'spreads', or identical standard deviations. These may be analysed using a measure of 'risk per unit return'. For example, two investments X and Y have identical standard deviations of £3,000, but $E(X) = £10\,000$ and $E(Y) = £50\,000$. 'Risk per unit return' will standardize the two distributions and show which is riskier. The measure is also known more conventionally as the coefficient of variation (CV).

$$CV = \frac{\text{standard deviation of returns}}{\text{expected return}}$$

$$CV(X) = \frac{3000}{10\,000} = 0.30$$

$$CV(Y) = \frac{3000}{50\,000} = 0.06$$

Investment X with the higher CV has a greater level of risk per unit of return than Y, five times as much in fact.

In these cases, the variance has been calculated around equal expected values, and the coefficients of variation around equal standard deviations, purely for demonstration purposes. The methods can and are applied generally for the assessment of risk when outcomes are probabilistic.

Decision trees

So far we have been looking at simple mathematical methods for evaluating the expected outcomes of decisions. Another form of analysis which is intended to clarify the possible courses of action that can be taken, and the consequences that will follow, is the decision tree.

A decision tree is a diagrammatic representation of a pay-off table. Decision tree analysis allows large or complex decision problems to be broken into smaller subproblems that can be solved separately and then recombined. It is most useful when the problem has some definable sequence.

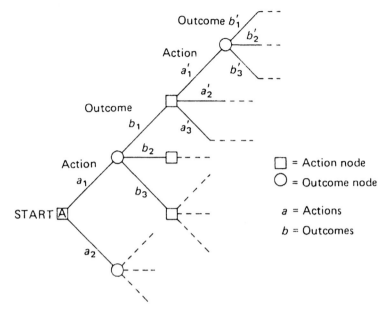

Figure 3.4 A decision tree

The basic structure of a decision tree is shown in Figure 3.4. Suppose a decision begins at A, with two possible courses of action (a_1 and a_2). If a_1 is selected then the possible outcomes are b_1, b_2 and b_3. Suppose that a_1 actually occurs. There is then the choice of three new courses of action a_{11}, a_{12} and a_{13} and so on. The action–outcome sequence expands from left to right in the diagram as a branching 'tree' representing the complete array of *all* possible acts and outcomes.

Branches split into further branches and the junctions are called nodes. They are of two types: those that are and those that are not within the decision-maker's control. The nodes labelled *a* are 'action' or decision nodes, and are usually denoted by a square on the diagram. These are the points at which the decision-maker has to select a branch from the choice available. Different courses of action may well be possible. The actions, in any decision are in principle under the control of the decision-maker. Even if coerced, it remains possible, if unwise, for a decision-maker to say 'no' to a particular course of action. The implication of this is that the more control the decision-maker can exert, the less uncertainty there should be in the system, or at least in some parts of the system. It does not follow, however, that controlled actions lead to 'safe' decisions.

Each action in turn leads to a set of possible outcomes (also called events, and labelled *b*). In the decision tree, the branch splits into outcomes at an event node, usually denoted by a circle. These outcome nodes are not, however, controlled, indeed they are the decision tree equivalents of states of nature, and may be assigned probabilities.

The tree is developed in three or four stages:

1. All possible action–outcome sequences are put on to the tree, working from left to right. Putting the decision problem in this form is not necessarily straightforward. Irrelevant elements need to be removed, so that only the bare branches of the problem remain. This is a question of seeing wood through the trees.
2. Numerical values which indicate the intermediate results then need to be evaluated together with the probabilities for the various uncertain outcomes. These are then put on to the diagram at the appropriate points.
3. The usual method of analysis is to 'roll-back' or 'fold-back' the tree working from right to left, reversing the process of tree construction. As each decision node is reached the action having the greatest profit or smallest loss is chosen, and rolled back to the next decision point.
4. A fourth stage, sometimes employed, is to apply sensitivity analysis to the tree, changing the variable values and probability distributions, and looking at the best courses of action which then result at each node.

As a simple example of what a decision tree might look like, the pay-off table from Table 3.3 is represented in its tree form in Figure 3.5. In the tree, the pay-offs are on the far right, representing final possible outcomes. Each of the outcome branches has the appropriate probability attached, and at the event node (the circles), the expected value of all the outcomes is presented. Thus for scheme A, the value here is −0.67. Thus, the tree is rolled back from right to left. The best of all the expected pay-offs is carried back to the decision node (the square). In this case this is 5.33, the expected pay-off for D. Assuming that the decision rule is EMV (maximize return), and there are no more stages in the decision, then the decision is invest in D.

Essentially a graphical method, the development of decision trees provides a first approximation for solving decision problems. Because of this, the process is as much an art as a science. The development of each branch and labelling actions and outcomes may stimulate new ideas as to how the problem may proceed. Each branch can be tried out, and if certain branches appear to dominate others, then the latter can be pruned out at an early stage so that the analysis is not confused.

This is really the purpose of the decision tree. Its great value is that it brings opinions into the open and organizes them into a sensible and structured framework. A finished, drafted, tree diagram need not be a final product, as long as a clearer understanding has been obtained of the way the problem may develop.

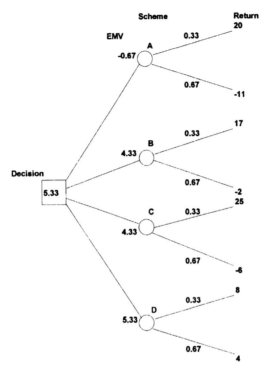

Figure 3.5 Decision tree structure of Table 3.3

3.3 UTILITY

The concept of utility

In some cases, Bayes' rule may indicate an action (outcome) which is nevertheless still felt to be unsatisfactory by the decision-maker. For example, consider a project which if it succeeds will give a gain of £200 000, but if it fails will give a loss of £100 000, the probability of success being 0.8. The EMV is therefore:

$$(0.8) \ (£200\,000) \ + \ (0.2) \ (-£100\,000) \ = \ £140\,000$$

Following Bayes' rule the EMV is positive and so the project is preferred to doing nothing, which is presumed to be the only alternative.

But suppose that the developer controls relatively small assets. In such a case a loss of £100 000 might be catastrophic whereas doing nothing might not have such serious consequences. The developer can 'afford' to forgo the opportunity of gaining £200 000 but simply cannot contemplate the loss of £100 000. Bayes' rule tells him to go ahead in what, for him, is clearly the wrong direction.

Suppose now that the developer is in the same overall financial state, but the pay-offs are only a tenth of those previously. In these circumstances, the developer may well be able to accept the risk of a loss of £10 000 and given the positive outcome may decide to proceed with the venture. This example shows that we have not necessarily gone far enough when we simply arrive at an expected outcome or EMV.

Another example of this is that of a householder buying a fire insurance policy. He is offered a policy at an annual premium. Even if the probability of a fire is so slight that the householder, adopting Bayes' rule, is really better off not buying fire insurance, most householders would do so, not because they think there is going to be a fire but because they prefer the security of the insurance policy at a modest annual premium to the unacceptable burden of having to replace their house at their own cost in the event of a fire, however improbable that is. We can say that the householder derives greater satisfaction or utility, in this case peace of mind, from having the insurance than from carrying the burden of the risk himself.

The use of the concept of utility enables monetary pay-offs to be replaced, if necessary, by a consistently appropriate measure of satisfaction, known as the util. Bayes' rule may then be modified to state that the best course of action is that which maximizes expected utility which, as we now see, can differ from expected monetary value.

Describing utility

There are a number of methods for measuring utility, but all methods lead to the production of a utility curve or utility function which describes the decision-maker's current attitude to risk. The utility curve is one of the best descriptions of an individual's attitude to risk or risk preference and different curves describe the three kinds of risk-taker who were identified in Chapter 2: the risk-seeker, the person who is indifferent to risk and the person who is risk averse.

It is unwise to attribute a general risk profile to any particular individual or corporation since looked at across the whole range of their attitudes over time – they move up and down the entire spectrum from risk averse to risk seeking – but for the purpose of illustration and related to our particular interest, one might say that the manager of a pension fund would tend to be risk averse whereas the individual entrepreneur or property developer would tend to be risk seeking. Utility curves in general are regarded as having a limited life and should be updated prior to each decision.

Types and characteristics of utility functions

The general characteristic of a utility curve is a consistent rise from left to right in the graph. Generally, most people ordinarily prefer more money to less. In economic terms there is positive marginal utility for money for most people.

Figure 3.6 shows three major utility curves. Curve A is characteristic of a risk averse individual. For most of his life the average person is risk averse; actions which involve high risk and/or the chance of larger monetary loss are avoided. The curve indicates a diminishing marginal utility for money as the monetary gain becomes larger, although marginal utility always remains positive. This is seen in the decreasing slope as the amount of money increases; as the money gain becomes larger its true worth to the individual becomes less.

Curve B is appropriate for a risk neutral individual. This individual is indifferent to risk, and each additional monetary gain or loss causes the

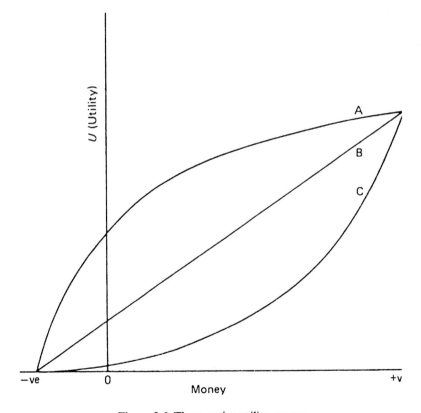

Figure 3.6 Three major utility curves

same utility increase or decrease. Such a view is apparently typical of the owners of enormous wealth, such as large corporations. These 'individuals' make decisions which are often based on Bayes' rule applied directly to expected monetary pay-offs, since their marginal utility is unchanged across the monetary scale, and is maximized by this approach.

Curve C is that for the risk-taker. Most individuals seek and take risks at some point in their lives, but the 'consistent' risk-taker will behave recklessly! This individual will see each successively gained monetary unit as more valuable than any gained previously. He is motivated by the possibility of achieving the maximum reward from even the most dangerous gamble; indeed the risk-seeker will prefer some gambles having a negative expected value to the status quo, since a possible large pay-off total outweighs the fact that the probability of that pay-off may be very small.

If only a small difference in money is involved, most individuals have utility curves which are approximately linear, where the slope does not change much. In business decisions, the assets devoted to any single project may be a quite small proportion of the total. In these conditions the utility curve may be of the risk neutral kind, and the use of expected monetary values may be quite justified, equating with expected utilities over part of the utility curve.

Looked at over the whole range of their utility, and over time, many individuals are naturally both risk averse and risk taking. This relates to levels of aspiration as they change over time, so that each of those general forms are the ideals for these kinds of risk-taker.

Friedman and Savage (1948) produced an example of a generalized utility curve for the individual which expresses both risk aversion and risk-taking circumstances, much more representative of the average attitude to risk. An individual, for whom this utility curve, shown in Figure 3.7, is appropriate, can take some risks (can gamble for instance) and can also avoid risks (by buying insurance of various kinds). Such an individual is quite reasonably risk averse and risk-taking at close points in time. Regardless of the probability of success for any outcome of a decision, utilities for those outcomes remain unchanged until a change is justified by the decision-maker himself. Utility curves in general are regarded as having a limited life and should be updated prior to each decision.

The assessment of utility

There are a number of methods for the assessment of utility. These methods all lead to the production of a utility curve or utility function which describes the decision-maker's current preference. All methods begin with the same basic stages:

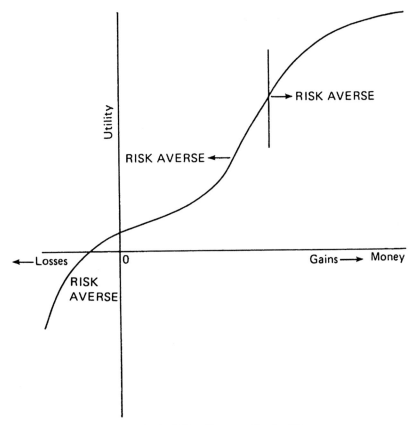

Figure 3.7 An individual's generalized utility curve

1. Decide the person or group to whom the curve is to apply. The scale of decision and factors involved will vary from level to level within an organization. Each decision level should have its own utility curves, even though they may be part of a 'greater' corporate utility function.
2. Decide the range of pay-offs within which the utility curve is to be operated. Within this range utility values have to be placed on all possible outcomes. These can be non-monetary, but in most business situations a monetary outcome is of greatest consequence, and we shall consider the evaluation of the utility function for money.
3. When the range of pay-offs has been decided, a maximum value of utility is equated to the highest pay-off (measured in its original units). The choice of utility scale is entirely arbitrary, but once selected it is best kept consistently for all future decisions with pay-offs within the range originally specified. The best known, and most

often applied method of assessing utility, is the method known as the 'standard gamble', first suggested by Von Neumann and Morganstern (1948), following the establishment of the set of rules outlined above. They said that as long as those rules are followed a utility curve can be tailored to the preferences of any 'individual'.

4. The method is in terms of a 'gamble' or 'lottery' approach. If outcome A is preferred to B which is in turn preferred to C, then there is some probability (p) such that

$$E(U(B)) = pU(A) + (1 - p)U(C)$$

This is, the utility of a gamble between two outcomes equals the expected utility of the gamble.

Systematic method

1. Rank all outcomes in order of preference.
2. Take the most preferred and least preferred as the two possible outcomes of the lottery (This is sometimes called the reference lottery, rather than the standard gamble, since all other utilities are calculated by reference to these two outcomes.). The lottery has two events:

   (a) Win i.e. achieve the best outcome.
   (b) Lose i.e. achieve the worst outcome.

   The probability (q) of winning the reference lottery changes, depending upon the pay-off with which it is compared, and on the decision-maker's attitude to that pay-off.
3. In order to establish the value of q for any pay-off against the reference lottery, use is often made of another standardized approach known as the 'standard urn'.

The decision-maker is asked to think of an urn in which there are a number (say 100) of counters or marbles. Some of these counters are labelled W for win, the rest are labelled L for lose. The decision-maker is asked to think of putting his hand into the urn and drawing one counter at random. If W is selected then the outcome (say) £200 will be assured, otherwise of course L must be selected and in this case a loss of say £200 will be the 'reward'.

The decision-maker is now asked to say how many of the total 100 counters would have to be labelled W before he thinks that he will *just* take part in the gamble. If he has already placed some arbitrary measure against £200, then the utility of −£200 to him can be calculated. The decision-maker says 75 win counters have to be in the pot. This means he wants to be 75% certain of winning 200 and 25% certain of losing 200 before he is indifferent between taking part in the lottery or doing nothing. It has been decided that 15 units of utility = 200 units of money. Then:

Expected utility of the gamble = $(0.75)0.15 + 0.25X$

but when $q = 0.75$, $E(U) = 0$ (doing nothing). Thus

$$(0.75)0.15 + 0.25X = 0$$

$$X = \frac{-0.75(15)}{0.25} = -45 \text{ utils}$$

This puts three points on this utility scale.

1. $200 = 15$ utils
2. $0 = 0$ utils
3. $-200 = -45$ utils

Any other value on the money scale can now be assessed in the same way, by reference only to the value of q, the win probability in the reference lottery.

The 'shape' of the reference lottery can be seen in the tree diagram (Figure 3.8). The reference lottery (branch A) has outcomes of +200 and −200, and is one action which the decision-maker can take. Alternatively the decision-maker can choose branch B, and take that pay-off for certain. This pay-off is sometimes called the 'certainty equivalent'.

In order to develop the utility curve in this case, the amounts given in Table 3.5 were offered, and the decision-maker was required to assess a q value for each as outlined above. Thus, when the pay-off offered for certain is −150, the decision-maker only wants to be 25% certain of winning 200 before preferring the lottery. At the other end of the scale, if

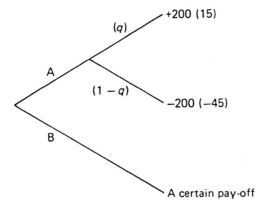

Figure 3.8 Shape of the reference lottery

Table 3.5 Information to assess a utility curve

Certain pay-off	No. of W balls in urn ($=q$)
−150	25
−50	60
20	80
80	90
190	99

the pay-off for certain is 190, then a pay-off of 200 from the lottery has to be almost certain to be obtained, in this case 99% certain. Given these values of q, the 'extreme' utility values previously calculated are used to derive intermediate utilities. For example, certain money pay-off = −150, $q = 0.25$:

$$E(U) = 0.25(15) + 0.75(-45) = -30$$

In this way a complete table of utilities can be devised (Table 3.6), and a curve drawn (Figure 3.9).

In general between 6 and 10 utils need to be assessed to enable a reasonable utility curve to be drawn. The value of q is simply a device to establish the difference levels. The probability selected for winning the reference lottery has *nothing* to do with the chance of the most favourable outcome occurring. In selecting the value of q the decision-maker has only expressed his attitude towards any outcome, as compared with the standard gamble.

A somewhat revised version of this procedure has been suggested (Coyle, 1972) to try to simplify and reduce the hypothetical nature of the reference lottery. This approach has the merit of coming closer to a true decision process.

The decision-maker is asked to consider a number of past projects which are typical of the kind of work done by him, such as past develop-

Table 3.6 Table of utilities

Certain pay-off	Utility
−150	−30
−50	−9
20	3
80	9
190	14.4

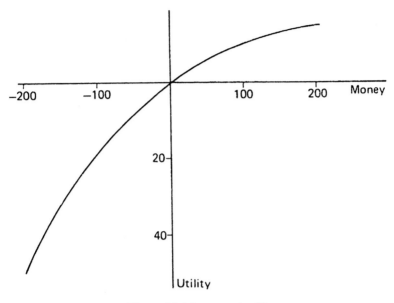

Figure 3.9 Money and utility

ments. They are converted to present values, and the decision-maker is asked to consider what chance of success each would have to have before they would be undertaken today. A utility curve can be drawn using the methods outlined above, for these past projects.

Outcomes from a *current* project can now be converted to utils on this curve and by a reversal of the previous process, a success probability can be calculated for this project, at which the utility curve implies indifference about undertaking the project. Suppose the project has pay-offs of $+200$ and -100 with utilities, taken from the graph, of $+15u$ and $-15u$ respectively (where u denotes utils). Then the indifference probability where $E(U) = 0$ is

$$p \times 15 + (100 - p) \times -15 = 0$$
$$15p - 15 \times 100 + 15p = 0$$
$$30p = 1500$$
$$p = 50\% = 0.5$$

The view taken is that the utility curve indicates that the project should be undertaken if the chance of success is greater than 50%. If the decision-maker's judgement affirms this, then the utility curve is considered correct. If not it can be adjusted until it is consistent.

If the decision-maker is in the end unable to adjust his choices he is acting inconsistently, and a utility curve cannot be constructed for him until the inconsistencies are removed.

Once the utilities have been derived, they can be applied to the modified Bayes' decision rule, maximizing expected utility. It is worth noting, however, that as long as a utility function is felt to be approximately linear over the range of outcomes in a decision problem, then the EMV version of Bayes' decision rule will hold, and will give the same 'best' decision, as if the utility measure had been used. Given the complexity of utility scoring, this is a worthwhile approximation!

The concept of utility and the development of the utility function, especially when taken together with the assessment of subjective probability, has two purposes. It provides a means of objectively describing the attitude of the individual decision-maker to risk as perceived at a particular time and it makes an important distinction between risk and uncertainty.

Uncertainty is measured by probability. When this is done subjectively what is in effect being measured is the extent of 'belief' in the occurrence of any particular possible outcome. In contrast, implicit in the process of measuring utility by reference to pay-offs is the principle that the decision-maker is looking independently at each of these pay-offs and is assessing an 'attitude' to each of them, expressly without reference to their probabilities of occurrence.

In property development and in the consideration of property investments generally, risk and attitudes to risk are poorly understood and are often wrongly interpreted, as is this difference between risk and uncertainty. Of the two the assessment of probability and its application is easier and more generally useful. It is only occasionally necessary to positively question attitudes to risk, but it is important to have in mind that systematic methods are available which make it possible so to do.

3.4 SENSITIVITY AND SIMULATION

Simulation, models and sensitivity analysis

All the techniques and approaches looked at so far in this chapter can be used separately to assist in various parts of the decision process. They can also be used to provide inputs to simulation methods. Simulation is one of the most powerful tools available for the analysis of business decisions, especially under conditions of uncertainty.

Using simulation methods, the decision-maker has a means of experimenting with the structure of a problem. By using a model of a problem, consequences can be tested without the worry of mistakes that might prove costly in practice.

Simulation is an analytical tool, to be used, like any other tool, only in appropriate cases. The development appraisal process is one area where, as will be demonstrated, a simulation approach can provide considerable benefits. This is because simulation provides a great deal of flexibility in the ways that it can be used. Many kinds of data can be put into simulation models. The models themselves can be constructed at various levels of detail and accuracy, and the model's output can be evaluated and modified in a variety of ways. This is especially useful for those problems which have many levels of uncertainty and for those which have many possible outcomes.

As will be shown, the use of simulation often involves extensive calculation. It is therefore an area in which availability of computers is of particular assistance. The computer should not be seen as in any way the problem-solver, but rather as the 'number-cruncher'. The simulation model has to be developed and then specified in a form appropriate for the computer. The necessary data have to be collected and prepared as input to the model, and once the computer has made the calculations, the output from the model has to be analysed and interpreted. The computer is only able to assist by performing one step in this process, and even then it is not essential, simply very useful.

Inevitably, the development of simulation models and the use of computers costs money. While the form of the model may be easy to conceptualize, the effort to turn it into an operational simulation model for the computer can be a relatively expensive process. The internal flexibility of these models can encourage the development of very extensive forms of model, where simpler, less expensive versions would still provide very satisfactory results. What is needed therefore is a clear initial specification or blueprint for the model, produced from a good understanding of the basic nature of the problem and from this, a comprehension of what simulation methods can do to assist in improving this understanding. These methods are clearly most appropriate for those problems where the risks are seen to be great, and where returns may be affected accordingly. They are also useful in unravelling very complex problems. In these circumstances the cost can be justified because the decision can be improved by the ability to consider alternatives properly. In any case, a sensible decision-maker will have considered the trade off between the cost of developing such models and the return to be obtained from their use.

Models and sensitivity analysis

Chapter 2 identified the various kinds of model that can be used by a decision-maker in the analysis of a decision. Although all kinds of model can be regarded as simulations, the simulations to be described and used here are all based upon the mathematical kind of model.

These mathematical simulations can be divided into two types:

1. Deterministic simulation models. A rather limited kind of model, in which all the variable factors input to the model either have values that are known for certain, or are assumed to be certain.
2. Probabilistic (or stochastic) simulation models. In these models uncertainty is treated explicitly, and any or all of the variable factors input to the model are represented not by known single values but are modelled as probability distributions.

Deterministic models

These primarily answer the question 'what if?' Simply, what happens to the results of the model if some or all of the known inputs are changed?

As we saw in Chapter 1, the conventional residual appraisal is essentially a deterministic model, although it is usually employed to answer a single 'what if?' question, rather than any repeated analysis. When a repeated analysis of even the simplest model is carried out it is called sensitivity analysis since the model's performance is assumed to be sensitive to changes in the input values. In fact any 'what if?' question should result in a sensitivity analysis, since simply asking the question implies that input values can change. If the decision-maker believes this to be so then a further question arises: 'How certain are the values which are being fed into this model as if they *are* certain?' In these circumstances, the deterministic model may form the first stage of a larger, and more valuable analysis, extending to probabilistic simulation, such as that shown in Chapter 5.

Probabilistic models

Where uncertainty is present, variables should be represented by probability distributions. Stochastic simulation makes the assumption that at least some of the variables in the model are described by distributions of this kind. The variables to be treated in this way will vary according to the particular problem. Some are likely to remain determined, others are represented by distributions. Variables which are described by probability distributions may be referred to as 'probabilistic', or alternatively, 'stochastic' or 'state' variables.

A full simulation, including many variables of both kinds, combines these variables in a way which expresses the expected performance of each, to produce an overall profile of the decision. The method, since it makes use of probabilities, can never provide an exact answer to any problem, instead it provides many rather inexact answers, which are then used to produce a range of 'most likely' figures for the decision-maker to interpret and act upon.

The Monte Carlo method

The most often employed procedure for including probabilities in a simulation is the Monte Carlo method, so called because it makes use of random numbers to select outcomes from the random variable, rather as the ball on a roulette wheel stops, theoretically at random, to pick the winning number. A roulette wheel could be used to select a sequence of random numbers, or dice could be thrown, both manual methods. Alternatively a computer can be used to generate a random number sequence. This has the advantage that it can be built directly into the complete simulation when it is carried out on the computer. Just how this can be done in the context of development appraisal will be demonstrated in much more detail in Chapters 5 and 6.

To demonstrate the process, a single variable can be simulated. The random variable in this case is the possible yield on an investment. The distribution is as given in Table 3.7.

If this distribution is plotted (Figure 3.10) the result is the symmetrical 'bell' shaped distribution, characteristic of the normal distribution. The shape of the distribution is dependent upon the probability of any value of this (random) variable being less than a particular yield.

What is being shown in Table 3.7 is that if the variable was to be repeatedly observed, say 100 times, a yield of 10% should be seen approximately 28 times (28% of the time), or that cumulatively there is a 64% probability that the yield will be 10% or less. The distribution is therefore a model of the variable's expected performance, if it could be observed repeatedly.

Suppose that on one trial the random number 27 is selected. If this number is inserted into the cumulative distribution in Table 3.7, it falls between 16 and 36. Any number between 17 and 36 is part of the prob-

Table 3.7 Random variable distribution

Yield	Probability	Cumulative probability ($\times$ 100)
6	0.01	1
7	0.05	6
8	0.10	16
9	0.20	36
10	0.28	64
11	0.20	84
12	0.10	94
13	0.05	99
14	0.01	100
	1.00	

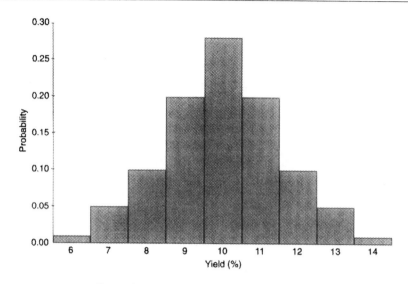

Figure 3.10 Distribution of investment yield

ability which has been attached to a yield of 9%. Hence the selection of the number 27 is equivalent to simulating the occurrence of a 9% yield. In the same way, selecting a random number between 37 and 64 would be modelling a 10% yield. The proportion of times that each of the possible values will occur is related to the cumulative probability distribution and thence to the shape of the probability distribution chosen to represent the variable.

This is a sampling process. Values of the variable are picked (sampled) from the distribution. Although the exercise is only repeated a limited number of times, this number is likely to be rather more than could ever be achieved by other means. One hundred is a reasonable number of repetitions.

When a computer is used, a special mathematical function is usually available to provide random sequences of numbers generally between zero and one, from a uniform (or equal probability) distribution. In this case, since the cumulative probability runs from 1 to 100, random numbers are generated between these limits. Each number generated has an equal and theoretically unbiased chance of being selected. Technically, computer generated random numbers are not truly random, and may be expected to repeat systematically after a very large series is drawn. This does not usually affect the 'apparent' randomness of the series, but should not be ignored.

If a count of each of the values of the variable is kept as they are simulated, a new frequency distribution will be derived. As the number of trials increases this distribution of values will look more and more like the distribution from which they are drawn. If this can be done a very large number of times the original distribution will eventually reappear, since this would be equivalent to simulating the long-run frequency (or probability) of the distribution (this is demonstrated in Chapter 6). Only the smaller samples usually taken prevent this from happening on any occasion when the distribution is simulated.

In a full simulation, the value selected at random from the variable is fed into the model, which will not ordinarily be confined to employing a single random variable. The same procedure is therefore repeated for every random variable in the model, sampling values from the distributions of each. This can be done manually, but is very applicable to computerization. The computer can store the cumulative distributions of each variable, can carry out the sampling, and store the values of the variables as they are sampled. Once this is done, the variables are combined in the model and any calculations are made and stored. The variables can then be sampled again. Different values of the variables will probably emerge to be combined and produce a new result to be stored until the cycle of simulations is complete.

The usual output from a computer simulation will be a profile of the full range of possible values measured against their relative probability of occurrence. The combination of the various random variables in the model is used to produce a new probability distribution of the results of the simulation.[4] This can in turn be used to make statements about the probable performance of the problem as a whole, such as the probability of making a loss. In addition the expected value, standard deviation and other statistics of the distribution are also calculated, so that its characteristics may be fully described.

3.5 CONCLUSION

This chapter has of necessity been theoretical, and 'technical' in ways which may be novel to many. The chapter has shown a set of quite basic methods for assessing uncertainty, and the rules for using them in a consistent way. By means of a discussion of utility, the differences in the ways that decision-makers view risk have been demonstrated. It is important to realize, though, that a regular assessment of utility is unlikely to be made in day-to-day decision-making.

Without a proper understanding of these concepts, a correct interpretation of risk and uncertainty in the context of a particular problem is not possible, and it is very unlikely that either will be properly accounted for in exercises like simulation, except by accident.

NOTES

1. After Thomas Bayes, an eighteenth-century Presbyterian minister and mathe-
matician, who developed ideas about probability and its assessment which
form the basis of much of modern decision analysis.
2. Figures 3.3(a) and (b) do not actually show a probability distribution, which
would be based on areas, but simply serve to illustrate the relative shapes of
the two distributions.
3. The variance is calculated in squared units to remove the negatives when x_i is
less than $E(X)$. To convert to normal units its square root is taken – the
standard deviation.
4. This distribution is often treated as a normal distribution, since this greatly
simplifies the analysis which can be carried out, but strictly this can only be
done if the variables which have been used in the simulation can be shown to
be independent, having no effect on one another. This is rarely the case.

Decision methods in development appraisal

4

4.1 INTRODUCTION

Chapter 2 demonstrated a straightforward, deterministic, approach to a development appraisal and, more importantly, identified the variable elements. On the whole the variables are quite easy to enumerate, but all are treated in the development appraisal as having an equal weight, and a single value. However, both subjectively and objectively this is not necessarily the case and attention should be given to means by which the sensitivity of the outcome to changes in the variables may be assessed.

Subjectively, the great difficulty is always deciding what the likely value of each variable will be. This depends in part upon whether the value of the variable is capable of being controlled by the decision-maker. If it is, then these variables may be regarded as fixed, although there may be good reasons for testing the performance of the model if these values are changed. This can obviously be done most easily if the appraisal is modelled with the help of the computer.

More attention must then be given to the uncontrollable variables in the model. Information about the likely performance of these variables may be gained from external sources, and should be sought, because under uncertain conditions part of the point of the exercise is to obtain and use the maximum amount of available information. Alternatively, using the methods outlined in Chapter 3, estimates may be made of the likely performance of these uncontrolled variables.

The sensitivity of the model may then be tested with both kinds of variable included, and the magnitude of the effects of change measured. In this fashion, the sensitivity analysis can precede any complete test of the model's performance, perhaps using simulation methods. Used in this way sensitivity analysis seeks to 'mark' those variables which will most affect the development during its life. This has a dual purpose:

1. To place these variables clearly before the decision-maker so that, as and if the development proceeds, these critical variables can

be particularly watched for changes which may affect overall viability.

2. To enable the values of the variables put into the model to be considered more carefully, so that as far as possible their values are 'best estimates' of their performance. This may be done in a variety of ways, but a standard sensitivity analysis only takes point (single) values for each of the variables at the start of the analysis and does not change these values during the analysis. Sometimes this does not matter, because it can be compensated for. For example, present building cost can be inflated by an amount which, in real life, will vary continuously. If the continuous variation is considered to be too complex to model, a workable alternative can be to attempt to provide an average (mean) figure for the inflation rate over the period of building.

However, the sophisticated analyst will seek to build such dynamic features into the system model. As a compromise this can be done by the assessment of probability distributions for at least some of these dynamic variables, for inclusion in the model.

Before considering how these dynamic models are used, it is worth discussing in more detail the ways in which probability assessments may themselves be used for analytical purposes.

4.2   ANALYSIS USING PROBABILITY DISTRIBUTIONS

Probabilities assessed by the kinds of method described in the previous chapter can be regarded as having two possible uses in the development appraisal process:

1. They can be assessed and used to add depth to an otherwise conventional analysis.
2. They can be used as input to a simulation model of any particular development.

The use of subjective probabilities

The assessment process itself enables consideration to be given to the range of possible values for the variables under consideration. For each of the values felt to be possible, probabilities can be assessed and a distribution constructed.

This procedure can be very valuable in its own right. Indeed, the analysis could quite easily be confined to an assessment of subjective probabilities since much information can be gained by a close consideration of the structure of the distributions, especially when this is related to the general attitudes to risk of those concerned.

The kind of analysis which then becomes possible can be demonstrated by reference to an actual example. In this case an attempt was made to use expert opinion and judgement to assess the probable performance of a particular development. Here the results from part of this experiment are examined. The development consisted of a phased warehouse development on an industrial estate, and for discussion purposes we now consider one element of that development, Unit A. Unit A was a warehouse of 33 000 ft^2 (3066 m^2). The costs, values and opinions shown here are those at the time the exercise was conducted.

For the exercise, which was being carried out conventionally for the client in any case, four variable factors were considered. These were:

1. Building cost
2. Building period
3. Rent
4. Letting period

For each of these factors, the appropriate expert adviser had produced a single estimate, and did not know at that time that he would be asked to attempt anything more complex. Later, however, these advisers were brought together and were asked to attempt to assess subjective probability distributions for the variable(s) against which they had previously put single values.

The basic method was outlined, and following the methods given in the previous chapter, the advisers proved able to produce workable distributions. It was interesting to see that those 'tested' later had learned the procedure by observing the earlier attempts, so that when their turn came the analysis could be carried out much more quickly and accurately.

(a) Building cost

The quantity surveyor's initial estimate of the Unit A building cost was £450 000. At the earliest stage in the assessment the quantity surveyor was unwilling to move from this value. He clearly felt that this was the price that the job would cost, although he was equally sure that the figure could and should be described as an estimate! There then followed a discussion with him about the definition of 'estimate', and eventually he conceded that other price outcomes were possible. He still felt that his initial estimate was 'best'. At that stage there was no further discussion about what 'best' actually meant, but it obviously had a particular meaning for the quantity surveyor.

The opening position occupied by this person is quite typical of an individual approaching a probability assessment of this kind for the first time. In particular there is a tendency to argue that 'words are being put into people's mouths'. As long as the procedure is carefully carried out this should not be the case, although gentle 'leading' may sometimes be required.

Eventually, by patient and careful questioning it proved possible to construct an equal fraction cumulative probability distribution to fit the quantity surveyor's expanded view of the way which he felt building costs might perform. From this a more workable cumulative probability and histogram were constructed (Figure 4.1). These show that there was, in the view of the quantity surveyor, a 45% chance that building costs could be less than £450 000.

Taking building cost in steps of £10 000, Table 4.1 shows the probability that cost would be less than each value.

Clearly the quantity surveyor believed that there would be a relatively

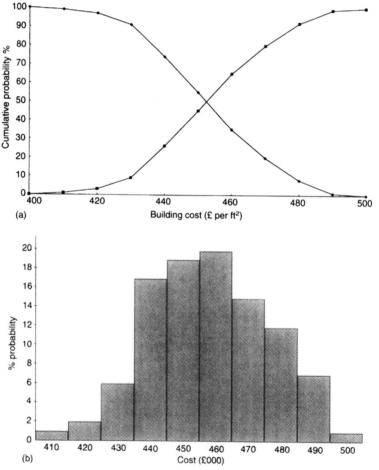

Figure 4.1 (a) Cumulative frequency graph of estimated building cost per square foot; (b) histogram of estimated building cost

Table 4.1 Probability of cost being less

Building cost (£ per ft²)		% cumulative probability
400	100	0
410	99	1
420	97	3
430	91	9
440	74	26
450	55	45
460	35	65
470	20	80
480	8	92
490	1	99
500	0	100

small cost overrun and has assigned most of his weight of opinion to values very close to his original cost figure. Overall there was a 55% chance that the original cost figure would be exceeded, but the quantity surveyor did not see the percentage overrun as being very great; he was, for example, 80% sure that cost would stay below £470 000.

In the case of the quantity surveyor this result seems reasonable, in the sense that he, more than the others concerned, would have access to large amounts of quite good information about the costs of the individual items making up his bill of quantities. He might be expected therefore to be reasonably confident about the final cost. However, seeing the results of this exercise, he became convinced he was unreasonable in initially feeling so absolutely certain, and from that point of view the exercise proved worthwhile. It also provided a much more considered view of the way that costs were likely to perform.

Building cost is closely related to the length of time taken by construction, and cost overruns are a very likely consequence of a time overrun. In some cases the level of building cost might be seen as conditioned by the length of time taken, and this might need to be allowed for in the analysis.

(b) Building period

In this case both the architect and quantity surveyor saw the building period as six months. However, following probability assessment and the construction of the appropriate distributions, a somewhat modified view needed to be taken (Figure 4.2). It was now seen to be rather unlikely (60%) that the building would take as long as six months, and the probabilities for a period longer than six months drop away rapidly. Less than

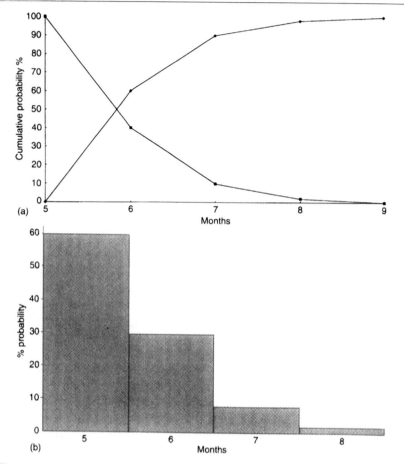

Figure 4.2 (a) Cumulative frequency graph of estimated building period; (b) histogram of estimated building period

six months was seen as virtually impossible, but a period of 5.25 months was felt as definitely possible. Slight savings might therefore be made.

Again with relatively good information about work schedules, etc., and considerable experience in handling this type of building, it was expected that the distribution for this variable would have a definite peak around the originally assessed value. This was clearly the case, but provided the valuable further information that the building period could be less than six months.

The letting agents were asked to consider the two remaining elements.

(c) Rent

Initially the agents were proposing a rent of £2.25 per ft², but as the probability distribution evolved it became clear that £2.25 represented some kind of upper limit, a rental level which was being aspired towards, rather than positively thought to be attainable. From Figure 4.3 and viewed against the proposed £2.25 rent there was a 62% chance that rent could actually be as low as £1.90, but the probability of its being lower than £2.00 was rather less than 10%.

Although centred around the initial estimate of £2.25, the overall distribution represents a rather pessimistic view of prospective rents. There

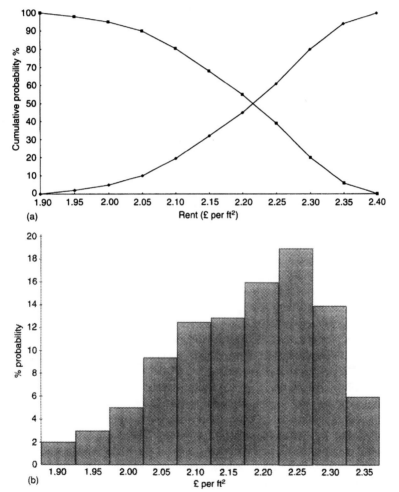

Figure 4.3 (a) Cumulative frequency graph of letting agent's estimated rental income; (b) histogram of estimated rental income

was according to the agent a 38% chance of achieving a rent greater than £2.25. Clearly therefore the agent favoured a rather conservative approach to the level at which the rent should be set. He would appear to have been attempting to 'guarantee' rental values up to £2.25, and to have been suggesting that anything above £2.25 was to be regarded as a bonus.

Against this expert opinion, understandably rather cautious, the decision-maker, having overall control of the development, would have to set his own attitude.

The level at which rents can be pitched is, in part, related to the point in the future at which they are first paid. While higher money rents might be expected further into the future, if only because of the level of inflation, set against this is the need to begin to make a return on capital, usually at the earliest stage. This may mean a lower rent is set in the hope that a relatively speedy letting will result, and income will flow.

(d) Letting period

The agent's first view of letting period was an assessment of the average time which the building would take to let. He felt that this was of the order of six months. In this case he was happy to offer an assessment of probabilities, since he was quite sure that the six month figure represented only one of a set of possible alternatives.

When the distribution is examined (Figure 4.4) this uncertainty can be seen. The distribution is rather uniform (flat) and demonstrates both pessimism and an element of optimism. 'Hedging his bets' the agent saw a 50:50 chance that the unit would be let in six months. However, he saw no chance of the letting taking less than three months. The 'tail' on this side of the six month point is three months, which can be viewed in an optimistic way, being relatively narrow. There is, though, a six month positive tail, with a 13% chance that the unit will remain unlet after nine months, and up to a year after completion.

It is perhaps unreasonable to describe the agent's probability analysis made at that time as pessimistic. It would be fairer to describe it rather as realistic, reflecting with a fair accuracy the agent's view of the prospective market for that kind of property.

The exercise discussed here was experimental, principally intended on this occasion for academic purposes. But equally, practically, these assessments made it possible to look at an otherwise traditional development appraisal in a somewhat different way. The appraisal calculations remained conventional, but discussions took place about the relationships between rental levels and letting periods, and on the possibility of reducing building cost and time.

Generally, this kind of analysis is a useful additional tool, intended to make explicit the uncertainty that underlies the so-called 'best estimate'

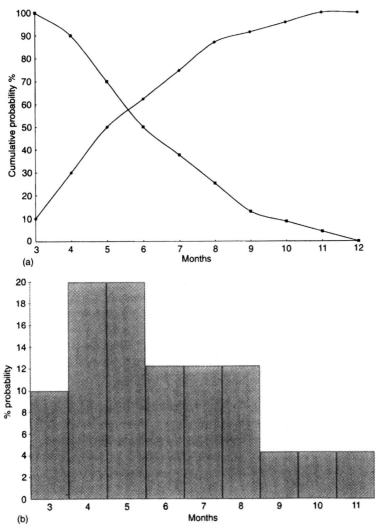

Figure 4.4 (a) Cumulative frequency graph of estimated letting period; (b) histogram of estimated letting period

and to enhance the judgement of the decision-maker and his team. In our experience, those introduced to this kind of analysis rapidly begin to see the advantages of this approach. For example, the distributions for the letting agent were derived last, and as this was to be done, the agent professed himself familiar with the basic idea of the analysis and was looking forward to showing himself capable of producing a meaningful distribution! By that stage he also saw the kind of additional information which would be available to him once the analysis was made. He also recognized that it was forcing him to consider and justify a full range of possibilities in the light of existing evidence and his experience. Most important, he had then to try to consistently 'weigh' his advice in the light of the probable outcomes which he had himself identified.

4.3   PROBABILITY DISTRIBUTIONS AND TIME

The principal reasons for deriving subjective probabilities are that no long-run evidence of any other kind is available, and that the variables under consideration will have values which cannot be known for certain.

The probable performance of these subjectively assessed variables may then be included in simulation methods, and this represents the second major application of these distributions. The subjective distributions then take their place with other, objectively, obtained distributions and with determined variables in the simulation.

There are various kinds of appraisal for which simulation can be used, and simulation models can be applied at varying levels of sophistication. However, there is one frequently modelled feature which can cause distributions to require modification or amplification before use. This feature is time.

Generally, uncertainty increases over time. Uncertainty is represented by the probability distribution, and any increase in uncertainty should be reflected by appropriate changes in the distribution of a particular variable.

This can mean a larger effort is required when assessing subjective probability. In the example given above the time horizon was quite limited. For the quantity surveyor it was until the development was completed; for the agent, up to twelve months after completion, when he proposed that it would be let. For some appraisals, however, the preview may have to be over a much longer time-base and adjustments have to be made to any probability distributions to allow for this. Subjectively, these adjustments would be made by requiring the expert to look further ahead in time, and produce distributions to match his view of what might occur. The result would be a 'family' of distributions describing the expected performance of the variable at each point in time.

On the assumption that uncertainty increases over time, the result might look like Figures 4.5 and 4.6. In Figure 4.5, an investment produces an expected cash flow (X) in period 1. The uncertainty around that expectation is in the 'spread' of distribution as a whole, measured by the standard deviation (σ). The investment also produces a cash flow in period 10. The expected value of that cash flow is assumed to be the same as in period 1, but the spread has been greatly increased, the distribution flattened, indicating a much greater uncertainty as to the actual, as

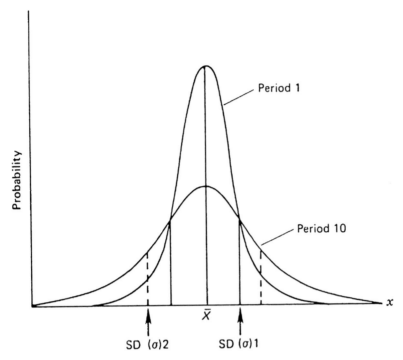

Figure 4.5 Increasing uncertainty reflected in a variable (*x*) by increasing variance over time

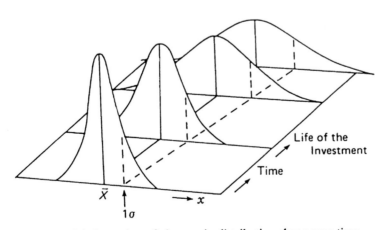

Figure 4.6 A number of changes in distribution shape over time

opposed to the expected, value of the cash flow in period 10. Over time the standard deviation and the variance will increase to reflect greater uncertainty. It may also be the case that the expected return will vary depending upon the symmetry of the subjectively derived distribution. This is not shown in these illustrations for the sake of clarity. This is also shown in Figure 4.6, demonstrating a number of changes in the shape of the distribution over the life of the investment.

Figure 4.6 implies that the variation is really increasing continuously. This may well be the case, and if so, a very large number of probability distributions would have to be constructed. This is clearly not realistic, and a compromise always has to be accepted, if the probabilistic variable is to be of use in the overall model.

It would normally be sufficient to represent measurable, or assessable, changes in the variable. This might require three or four distributions for the variable in an appraisal of twenty-four or thirty-six periods. This approach to time should be applied to all probabilistic variables in the simulation.

A further correction should be applied to the probabilistic variables, a feature which has been mentioned as part of the subjective assessment process. Each variable is assumed to be independent of all others. As indicated in the examples above this is very often not the case. If variables are dependent on each other it is formally necessary to develop conditional probabilities to account for the interdependencies. This can be a difficult and time-consuming process and is often ignored in simulation exercises.

However, the way in which such interdependencies might be structured is shown below. A probability distribution for construction time might be as in Table 4.2. Against this may be set a series of probability distributions of construction cost, the underlying assumption being that as time passes, cost probably increases.

Table 4.2 Construction time (months)

Months	3	4	5	6	7	8	9
Probability	0.05	0.10	0.15	0.20	0.25	0.20	0.05
	A		B		C		
Construction costs	Cost	Prob.	Cost	Prob.	Cost	Prob.	
	400	0.10					
	410	0.10	410	0.05			
	420	0.15	420	0.10	420	0.02	
	430	0.25	430	0.15	430	0.03	
	440	0.40	440	0.30	440	0.05	
			450	0.25	450	0.30	
			460	0.15	460	0.50	
					470	0.10	

If construction time is three to four months, likely costs are represented by A. If construction time is five to six months, likely costs are distributed as in B. Should construction time be as much as seven, eight or nine months, costs are likely to rise substantially, and this is reflected in the distribution C.

Evaluating uncertain cash flows over time

A conventional, deterministic, discounted cash flow (DCF) does not, of course, take any account of the risk and uncertainty inherent in the assessment of returns being made at periods in the future, except that the discount rate often may be 'adjusted' to compensate for uncertainty.

The rate of return used in a DCF should be a riskless rate of return to find a net present value which is adjusted for the time value of money, but is not adjusted for risk. There are sound theoretical reasons why the discount rate should not be risk-adjusted (Bierman and Smidt, 1988). A deterministic DCF usually has the form shown in Table 4.3, although there are variations.

Assume an outlay of 100 initially, with returns over the following three periods. The riskless discount rate is 6%. Here the assumption is that *all* revenues are known for certain over the life of the investment.

Table 4.3 Deterministic DCF

Period	Cost/revenue	Discount factor @ 6%	Discounted value
0	−100	−	−
1	50	0.9433	47.17
2	20	0.8899	17.80
3	−40	0.8396	−33.58
		Present Value =	31.39

Revenues are 31.39. Costs are 100. Net present value = 100 − 31.39 = −68.61. Costs exceed revenues by 68.61, therefore the investment is not profitable.

In the case of a property development, however, some of the cash flows will be known for certain, but by no means all. There are some general approaches which have been developed for dealing with such cases.

The most straightforward methods for the evaluation of uncertain cash flows require two assumptions to be satisfied. First, the expected returns for each investment period must be normally distributed (see the appendix). Second, there is also an assumption that the cash flow(s) in any one period are independent of cash flows in any other period (Technically they are uncorrelated.) The method can be extended to cover cases where this is not so, but this is not discussed here.

Given those assumptions, the net present value (NPV) of an uncertain stream of returns can be found with the use of the following equations (see Hillier, 1963):

$$PV = \sum_{t=1}^{n} \frac{E(R)_t}{(1 + k)^t} \tag{4.1}$$

$$\sigma_{pv} = \sqrt{\sum_{t=1}^{n} \frac{\sigma_t^2}{(1 + k)^{2t}}} \tag{4.2}$$

Here:

$E(R)_t$ = expected return from the investment in the tth period, the probability weighted average

σ_t = standard deviation of the expected returns in the tth year

PV = PV of all expected returns over the n-year life of the investment

K = the appropriate rate of discount for future returns – in this analysis the riskless rate

σ_{PV} = the standard deviation of the PV of the expected returns

To analyse this cash flow, we look at the project, *as a whole*, taking the sum of the parts. There are several ways by which this can be done. For example, consider two projects, A and B. We will look at these projects individually to show the ways in which uncertainty may enter the analysis. Generally:

1. Both investments involve an outlay of 100 at $t = 0$.
2. Returns are expected over three periods.
3. There are five possible states of nature (market conditions, for example).
4. The discount rate (k) is, in this case, 10%.

Project A

Details are shown in Table 4.4. Notice that in period 2 the probability distribution ($P2$) is flatter than in period 1 ($P1$), and in period 3 ($P3$) it is flatter still, and skewed left – towards the possibility of lower returns. This gives an indication of a greater uncertainty associated with returns in the further future. The calculation of the standard deviation for period 1 is shown in Table 4.5.

$$\sigma_1^2 = \sum_{t=1}^{5} (CF_1 - E(R))^2 P_1 = 120, \ \sigma_1 = 10.95$$

Table 4.4 Project A

State of nature	Period 1 Cash flow			Period 2 Cash flow			Period 3 Cash flow		
	1	P1	(CF1P1)	2	P2	(Cf2P2)	3	P3	(CF3P3)
1	50	0.1	5	20	0.1	2	−40	0.1	−4
2	60	0.2	12	40	0.25	10	30	0.3	9
3	70	0.4	28	60	0.3	18	50	0.3	15
4	80	0.2	16	80	0.25	20	80	0.2	16
5	90	0.1	9	100	0.1	10	140	0.1	14
Expected values	$E(R)$		70			60			50

Table 4.5 Standard deviation for period 1

State of nature	Cash flow 1	$E(R)$	P1	$(Cf1-E(R))^2 P1$
1	50	70	0.1	40
2	60	70	0.2	20
3	70	70	0.4	0
4	80	70	0.2	20
5	90	70	0.1	40
			Variance =	120

Similarly for periods 2 and 3:

$$\sigma_2^2 = 520, \ \sigma_2 = 22.80$$

$$\sigma_3^2 = 1920, \ \sigma_3 = 43.82$$

Applying Equations (4.1) and (4.2) above:

$$PV = \frac{70}{1.10} + \frac{60}{(1.10)^2} + \frac{50}{(1.10)^3}$$

$$= \frac{70}{1.100} + \frac{60}{1.210} + \frac{50}{1.331} = 150.79$$

$$\sigma_{PV} = \sqrt{\frac{120}{(1.10)^2} + \frac{520}{(1.10)^4} + \frac{1920}{(1.10)^6}}$$

$$\sqrt{(99.17 + 355.17 + 1083.79)} = \sqrt{1538.13} = 39.22$$

Project B

In this case, the probabilities remain the same for the entire pay-off period, but the expected returns drop with each successive time period, as shown in Table 4.6. In this case there can only be one σ_t

$$\sigma_t = (0.10(20)^2 + 0.20(10)^2 + 0.20(10)^2 + 0.10(20)^2)^{1/2}$$
$$= \sqrt{120} = 10.95$$

Notice though that in project B, the coefficient of variation increases to period 3 (Table 4.7).

Here we can see the riskiness of B is increasing over time. Then:

$$PV = \frac{60}{1.10} + \frac{50}{(1.10)^2} + \frac{40}{(1.10)^3} = 125.92$$

$$\sigma^2 = \frac{120}{(1.10)^2} + \frac{120}{(1.10)^4} + \frac{120}{(1.10)^6} = 99.17 + 81.96 + 67.74$$

$$= \sqrt{248.87} = 15.78$$

Table 4.6 Declining expected returns

		Period		
State of nature	P	1	2 Returns	3
1	0.1	40	30	20
2	0.2	50	40	30
3	0.4	60	50	40
4	0.2	70	60	50
5	0.1	80	70	60
	E(R)	60	50	40

Table 4.7 Coefficient of variation in project B

1	2	3
10.95	10.95	10.95
60	50	40
0.18	0.22	0.27

Now we know that the costs of each investment are 100 at period 0. The NPVs are therefore PV – cost:

project A: 150.79 – 100 = 50.79
project B: 125.92 – 100 = 25.92

On the assumption that the net present values are distributed normally, which will be so because the returns are individually normally distributed, the probability of achieving a positive NPV can be calculated, for each of the projects. As a first step the normal deviate (Z) (see the appendix) has to be derived for the case shown in Figure 4.7. The area to be derived is that to the right of point A where NPV = 0 since this is the probability that NPV will be positive.

Then the normal deviate is given by:

$$Z = \frac{PV}{\sigma_{NPV}}$$

σ_{NPV} can be taken to be equal to σ_{PV} since the difference is a constant:

$$Z(A) = \frac{50.79}{39.22} = 1.30$$

Using the tables available to give the area under a normal curve, the probability that NPV will be greater than or equal to zero is found to be 0.9032 for project A. The probability that the NPV will be negative is then (1 – 0.9032) = 0.0968, nearly 10%.

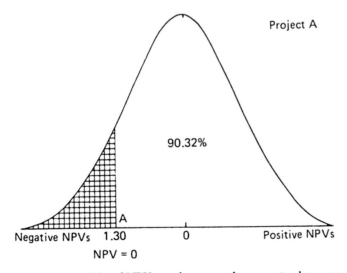

Figure 4.7 Area to the right of NPV equal to zero; the present values are normally distributed

Following this process for project B also:

$$Z(B) = \frac{25.92}{15.78} = 1.64$$

giving a probability of 0.9495. The loss probability here is $(1 - 0.9495) = 0.0505$, just over 5%.

This means that project A has a 1 in 10 (approximately) chance of making a loss, while for project B that chance is 1 in 20, making that project less risky. Against this of course the average return on B is only half that of A, so that the decision as to which to proceed with will depend upon the decision-maker's attitude to risk.

Other methods

There are other more extensive methods for dealing with uncertainty and risk in investments generally, but which can apply quite directly to property as an investment, and of course any development can be viewed in this way. These methods are too advanced to be treated in this book, but interested readers should refer to Hillier (1963) and Hertz (1964).

4.4 INFORMED APPROACHES TO DEVELOPMENT APPRAISAL

The conventional development appraisal, or residual valuation, can be criticized for a number of reasons, not least the inherent imprecision of the input values of the variables. In what follows we seek to identify an appraisal methodology which, while not entirely removing these difficulties, goes a considerable way towards minimizing their effects.

The first step, as outlined in Chapter 1 with the most basic residual methodology, is the construction of a computerized version of the model. In that simple model, a number of deficiencies were identified, which are not accounted for. This being so, what advantages, if any, can be claimed for the use of such a model? While in theory subject to many reservations, computerization of even this simple model makes the calculations much more consistent, accurate and straightforward, as long as the basic model remains acceptable on other counts. Next, the model is capable of demonstrating the effects of some kinds of variability, subject to very rigorous constraints, so that it can then be used for a limited set of sensitivity tests. The most rigorous of these constraints is the rigidity of the traditional appraisal, our basic model.

All the inputs to this appraisal model are fixed from the start of the analysis and the structure of the model does not permit ease of change, especially changes of the dynamic sort mentioned above. Sensitivity tests are therefore equally restricted, in that the kind of changes that might be expected to occur during the development cannot be realistically modelled.

The basic appraisal model can be improved in a limited way, however, to the extent that the spreadsheet can be constructed in such a way that values input to the model may be easily adjusted and the model's sensitivity can be tested.

Additionally, using the computer, some of the variables identified in Chapter 1 may be assessed to high accuracy (with minimum error). For example, the percentage elements accounting for agent's fees, legal costs, professional fees, etc., can be accurately calculated in every case. Even though they may represent only small proportions of total development costs there can be no case for making calculations less (or more) accurate than they need be. Once such adjustments have been built into the model, sensitivity analysis may be attempted. Using the Chapter 1 development appraisal as the no change case, Table 4.8 shows the separate effect of:

1. Varying building period by ±10%
2. Varying the short-term rate of interest by ±10%
3. Varying in the building cost by ±10%
4. Varying the yield and by ±10%
5. Varying rent by ±10%

All the other variables keep their original values.

Important elements of this table can also be presented in the form of a composite sensitivity graph, also sometimes called a spider diagram, showing the changes in the critical variables as percentage deviations away from the originally determined value. The form of such a graph is shown in Figures 4.8 and 4.9. In Figure 4.8, the effects of the variables are expressed in terms of developer's yield, while in Figure 4.9 the same variables are shown against developer's profit. Sometimes this graph is presented with the axes reversed, but the principle remains the same. In this graph, a variable which is horizontal, or close to it, has a small influence upon the outcome variable, while those with large angles and long lines have most effect. Thus, in Figure 4.8 the variables building cost and rent are seen to have most effect on developer's yield, but in opposite and partly cancelling directions, while developer's profit (Figure 4.9) is affected most by the same variables, plus a strong impact from the size of the investment yield. The range over which the sensitivity can be tested is of course alterable, but ±10% is often used as a benchmark level of variation for sensitivity analysis.

Scenarios

An extension of sensitivity analysis which has found some favour is the idea of the 'Scenario' or 'Scenarios'. A scenario is a set of variable values which model a particular decision situation. Three combinations are often examined for this kind of analysis, a pessimistic situation, the most likely,

Table 4.8 Effects of change of –10%, 0 and +10%

	–10%	0%	+10%
Building period			
Total development cost (£)	2695 868	2724 151	2752 803
Rental income (£p.a.)	250 000	250 000	250 000
Developer's yield (%)	9.27	9.18	9.08
Capital value (£)	3571 429	3571 429	3571 429
Developer's profit (%)	32.48	31.1	29.74
Short-term interest rate			
Total development cost (£)	2696 333	2724 151	2757 929
Rental income (£p.a.)	250 000	250 000	250 000
Developer's yield (%)	9.29	9.18	9.06
Capital value (£)	3571 429	3571 429	3571 429
Developer's profit (%)	32.75	31.1	29.5
Building cost			
Total development cost (£)	2518 957	2724 151	2929 345
Rental income (£p.a.)	250 000	250 000	250 000
Developer's yield (%)	9.92	9.18	8.53
Capital value (£)	3571 429	3571 429	3571 429
Developer's profit (%)	41.78	31.1	21.92
Yield			
Total development cost (£)	2724 151	2724 151	2724 151
Rental income (£p.a.)	250 000	250 000	250 000
Developer's yield (%)	9.18	9.18	9.18
Capital value (£)	3968 254	3571 429	3246 753
Developer's profit (%)	45.67	31.1	19.18
Rent			
Total development cost (£)	2720 401	2724 151	2727 901
Rental income (£p.a.)	225 000	250 000	275 000
Developer's yield (%)	8.27	9.18	10.08
Capital value (£)	3214 286	3571 429	3928 571
Developer's profit (%)	18.15	31.1	44.01

where the determined or expected (in the non-statistical sense) case might be taken, and the best or optimistic case. It need not be the case that the pessimistic scenario is the worst possible case, only that it represents a set of circumstances which are poor relative to the expected situation. Similarly the optimistic case need not be the best possible of all variable combinations, merely one which is clearly superior in pay-off terms than the

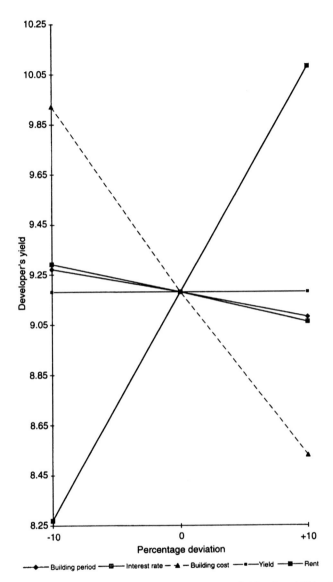

Figure 4.8 Sensitivity graph: effect on developer's yield of a 10% change in variables

expected. The values for input to each of the scenarios are chosen by the decision-maker or his advisers. The value of this analysis really lies in showing the decision-maker that the effect of quite small changes in one particularly sensitive variable can be dramatic, and in the pessimistic case,

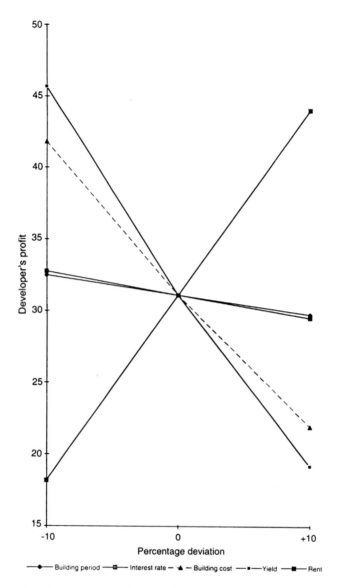

Figure 4.9 Sensitivity graph: effect on developer's profit of a 10% change in variables

dramatically bad! Table 4.9 shows the results of a typical scenario analysis.

Using our Chapter 1 example again, the values used here reflect good and bad combinations of these variables that could arise, compared with

Table 4.9 A typical scenario analysis

| Variable | Scenario | | |
	Optimistic	Original	Pessimistic
Land asking price (£)	400 000	500 000	550 000
Finance rate (%)	13	17	19
Building cost (£ per ft^2)	13.5	16	18.5
Rental income (£ per ft^2)	3.5	2.5	2.5
Yield (%)	6	7	8
Developer's yield (%)	15.72	9.18	7.93
Developer's profit (%)	161.99	31.1	−0.89

the original combination, which is assumed to be the most likely in this case. The variation that results is shown as substantial in both directions. It is also very sensitive. The pessimistic case here gives a negative developer's profit. Thus for example, a further reduction of £0.1 in rent per square foot leads to an additional 4% reduction in overall developer's profit, and each successive reduction by this amount would diminish the percentage profit by a slowly growing amount.

Although this approach is intuitively acceptable, it suffers from several technical shortcomings. First, sets of good and bad combinations of variables have been put together in ways which are very unlikely to occur in reality. These combinations could happen, but our knowledge of probability might suggest to us that the likelihood of it happening is really rather small. Equally it is assumed in the scenario analysis that the states of nature giving rise to these three combinations are equally likely. Again this may be so, but we should recognize that the states of nature need not be equally likely, and the probabilities which might be attached to them could colour our decision-maker's attitude to the outcomes. Thus, what is the probability of our original, most likely, scenario? Is it much more likely than either of the other two, and in particular, how high a probability should be given to the pessimistic combination? Although it is possible to modify the basic scenario method to cope with probabilities for these states of nature, a much more satisfactory and effective method is to estimate the likely values of the variables in each scenario, and assess whether they should be viewed individually in probability terms, as inputs to a simulation type model of the development appraisal.

The almost ubiquitous availability of computers now means that much less restrictive and more extensive models of development appraisal including sensitivity and scenario types of analysis can be used with relatively little computing expertise on the part of the average professional user. Many of these models come as part of 'packaged' development appraisal software, and a number of these packages with varying levels of sophistication are now available.

These packages all include a feature which removes one of the most unrealistic assumptions in the basic model, that is that costs and returns are spread evenly through the life of the project. Using this assumption can result in very inaccurate appraisals, or more charitably, appraisals that produce very conservative (low risk) results. A much more flexible approach to development appraisal, used in software packages, is one in which the inflows and outflows of cash are modelled and calculated on a period by period basis, and profitability is assessed by discounted cash flow methods. This is so much the case that DCF methods are now essentially the standard tool for development appraisal.

Although there are a number of detailed cash flow approaches which can be adopted (see Baum, (1978), for example), all are based upon much more extensive pre-assessment of patterns of expenditure and income, so that the model of the development process realistically represents the specific problem.

This approach has a number of general advantages:

1. Changes in costs and prices can be built into the model.
2. The timing of outgoings and returns can be accurately, and truly, represented. They can be adjusted period by period to test the sensitivity of overall returns to such changes.
3. Finance costs can be accurately computed.
4. The time structure of the problem, elements such as phasing, or partial rents received, for example, may be included in the model.
5. Much more information becomes available to the decision-maker, improving the decision.

The construction of a model of the development process making use of such methods is more complicated than the kind previously demonstrated, simply because it has to have far more flexibility, and hence constraining features have to be minimized. Such flexibility has the added advantage of greatly extending the range of sensitivity analyses which can be carried out and, as will be shown, makes possible the use of the simulation methods outlined in Chapter 3.

Equally, modelling based on these methods imposes a requirement for a much more careful consideration of the real structure of many of the variables, not only their values at the start of the development, but also their incidence throughout the development, along with any changes that are thought possible. This kind of data collection exercise is not worthwhile for a 'traditional' appraisal where the input values are assessed in the most basic way. It becomes essential for a detailed cash flow analysis.

A considerably revised view may now be taken of the way in which the variables work together in the model, and a more extensive consideration given to the shortcomings of the traditional method, looking particularly at the critical variables.

(a) Asking price for the land

Any development appraisal can be viewed in two ways:

1. For a given profit level (or percentage return) a maximum price to be paid for the land can be determined. In this instance the land cost is entered as zero, and the cash flow analysis will yield a present value, positive if the project is viable. Alternatively a 'bid' or suggested price might be entered as a test and a positive percentage return or present value looked for.
2. Where the purchase price is known, or the land is already in ownership and is input at a fixed figure, the appraisal results in a solution similar to that in Chapter 1, that is, a percentage return.

In either case, in a cash flow analysis, the land price usually appears as a cost (net outflow) early in the development as do any incidental costs incurred in acquiring the land.

The way in which the land can be procured varies from appraisal to appraisal, and might even be spread over a number of periods. The ability to do this can result in savings over the whole development, remembering that interest on any funded money is paid only in relation to the proportion of that money borrowed in any time period.

(b) Building costs

In the traditional appraisal this variable is measured in a very crude way (e.g. $100\,000$ ft^2 × £16 per ft^2, including the cost of roads, services and landscaping). The most unrealistic aspect of this variable is that the cost is spread evenly over the twelve month construction period.

In practice this is never the case. Early costs under this heading are associated with site preparation and infrastructure provision. Full construction will not begin until this is largely complete. Often construction costs rise to a peak soon after the middle of the total development period and then decline to the end of the period. A typical development cost pattern is shown in Figure 4.10.

In any case, costs of this kind are very likely to be subject to changes due to inflation over the life of the development. Even in a highly evolved analysis it is simplest to allow the costs to be assessed at current values, and then to impose appropriate inflation rates over the construction period. The inflation rate might well be viewed as a stochastic variable.

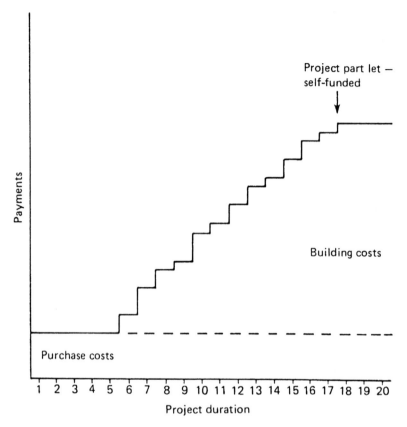

Figure 4.10 Construction cost profile (building starting in month 6)

(c) Professional fees

The professional fees associated with construction are around 14% of the overall cost, but as with building cost payments are not paid as a lump sum. They are spread over the construction period in ways which reflect the actual burdens of work encountered by the architects, engineers and surveyors, usually in arrears of the work being done on site, again important when money is borrowed.

These costs would also be inflated at the rates appropriate to the other construction costs.

(d) Rental income

The primary cash inflow is derived from rental income and/or sale proceeds. Rental income, in particular, is going to govern not only the value of the completed project, but is also likely to affect the overall viability of the development. On a cash flow basis, the sooner that rental income can appear, the higher will be the overall rate of return on that cash flow, since any remaining borrowing can be offset in the net cash flow.

The appraisal as a whole is therefore likely to be sensitive to changes in the patterns of rental income. In contrast, sale proceeds, although relatively massive overall, may have much less of an effect on the development as a whole, usually coming very close to the end of the payback period when little borrowing remains to be offset.

When rental income is present, the rental income currently assessed is likely to undergo changes of an inflationary kind between the start of the project and the first rental inflow, in the same way as building costs or may be negotiated to a different value.

(e) Letting costs

The advertising cost (£5000 in this example) is a subjective variable which would be decided by the developer and his agent in the light of market conditions. In a cash flow analysis, however, it will be seen that this outflow can be 'placed' within the cash flow much more carefully than in the conventional appraisal so that the net effect is minimized if possible.

In contrast this is less likely to be possible with fees to the letting agent, although it is fairly certain to be associated with the pattern of rental income, and is unlikely to be paid until any letting has been achieved.

(f) Interest payments

Interest payments form the third major variable component in an analysis. They are usually an outflow, but this is not invariably the case.

Interest will be paid on the net land and construction costs, and the amount paid will depend upon the amount borrowed in any period. One of the advantages of a cash flow (or period by period) analysis is, that just as construction cost cash flows can be distributed realistically throughout the development period, so also can interest charges.

This is clearly not the case in the present example. The figure of 17% used here reflects current and future possible rates in a single figure. This single figure is also intended to account for uncertainty, and the use of the 'approximation' $(12/2) + 3$ months is intended to represent the spread of interest charged during the development. This makes proper testing for sensitivity to interest rate changes rather difficult except in a very global way.

In the cash flow model, interest charging options need to be flexible if the sensitivity of the problem is to be tested against various patterns of interest payment. This can be especially important if probability distributions are to be employed.

Care should be taken in any calculation of interest to ensure that the effective rather than the nominal rate is taken. A sophisticated model may take a currently applicable annual rate and apply compound periods, which are likely to be at least half-yearly and may be more frequent still. Additionally, the current rate can also be inflated or deflated by amounts which are considered to be appropriate, given a prediction of the state of the money market over the life of the development.

4.5 CONCLUSION

The development of an extensive methodology for the analysis of decisions has made available the kinds of technique discussed in this chapter.

The assessment of subjective probability distributions makes the decision-maker aware of areas of uncertainty which his advisers have recognized, and equally encourages those advisers to include this recognition in the kinds, and quality, of advice that they offer to the decision-maker. This will also have an impact on the ways that this advice can be integrated into the decision process. For example, the decision-maker and/ or his advisers may perhaps resort to the decision tree to assist in the analysis of uncertain alternatives. Possible outcomes, however unlikely, are identified, discussed, perhaps rejected, and the entire viable structure of the problem over its lifetime is constructed. Even when uncertainty is allowed for, the graphical method of decision trees often demonstrates the way that a problem becomes progressively more unmanageable into the future, when viewed from the present. Additionally, the financial uncertainties have to be treated rather differently. The difficulty in this case is that costs and revenues have to be discounted to allow not only for the time value of money, but also for the time value of uncertainty, both of which can be regarded as being 'worth' less in the future.

At each stage, and in each process, the purpose is the minimizing of uncertainty, and by extension an increase in the amount of information as to the overall riskiness of the decision.

Using detailed appraisal models, this uncertainty can be allowed for using well tried capital budgeting methods, modified to account for uncertainty and the time value of money.

Modelling of the financial aspects of the development process with this degree of complexity is expensive and time consuming. It is not cost-effective except by using computers. The availability of computers on a wide scale does now mean, however, that much more flexible and extensive models of development appraisal can be employed through the use of any of the number of 'packaged' programs which are now available in the property software market or even by using spreadsheets.

Development appraisal the Monte Carlo way | 5

5.1 INTRODUCTION

Chapters 3 and 4 examined parts of the appraisal process. Emphasis was laid on the points at which the inclusion and assessment of uncertainty and/or risk would make a difference to the way in which the analysis is carried out. In particular Chapter 4 developed the idea of deriving probabilities subjectively for at least some of the variables commonly seen in development appraisals.

The principal means by which these elements may be utilized in a practical way is by applying simulation, using in this case the Monte Carlo method, outlined in Chapter 3, to the appraisal process. Thus, in this chapter, a single appraisal example is used to demonstrate various aspects of this approach in more detail. For this purpose a computerized model of typical appraisal has been developed.

A development appraisal is a standard spreadsheet application, and a modern spreadsheet package, associated with a reasonably powerful PC, can cope easily with the calculations involved in any development appraisal. Indeed, the sophistication of spreadsheets is such that it is now entirely possible to construct simulation models within the spreadsheet which are capable of dealing with most if not all of the technicalities which arise in such models. The spreadsheet used here is version 5 of Microsoft's EXCEL.

While the discussion which follows attempts to provide a fairly full and rigorous analysis of the problem, it concentrates on describing the form that a simulation approach will take, highlighting the important elements and calculations inherent in the method. Even so, the model is greatly simplified in some respects but a number of the input variables are described by probability distributions.

5.2   CASE STUDY: GENERAL DISCUSSION

The cash flow method was shown in the previous chapter to have a number of distinct advantages over the traditional approach to development appraisal. It is for those same reasons that the method is very suitable for adaptation to Monte Carlo simulation. These features are demonstrated by reference to the following example. This is a residential development, but the principles are essentially the same for any other kind of property development or investment applications.

A two hectare site has planning permission for the construction of twenty-four houses. These would sell now for £100 000 each, but prices are rising at 5% per annum. Initial site work will take three months, and will cost £45 000. Building will be phased over eighteen months, starting in month 4. Two houses will be completed and available for sale in each month after month 12. Building costs are currently £50 000 per unit but have been rising at 8% per annum. The developer requires an allowance of 15% of gross development value (GDV) for profit and contingencies. Architects' and quantity surveyors' fees are £150 000, payable in months 10 and 21. Agent's fees on sale are 3% of sale price and 3% on site acquisition. Finance can currently be obtained at 1% per month on the outstanding balance (12.68% per annum).

Rather than providing values for developer's profit and yield, this analysis is designed to produce a residual amount for the land. In some cases the asking price for the land may already be known, but if the purchase is competitive, knowledge of the maximum possible bid price may provide a competitive edge, or may ensure that a loss-making bid is not inadvertently attempted. Taking these values as determined the following calculations can be made.

(a) Income

Estimated sale price per unit in month 12 £100 000 $\times$ 1.05 = £105 000
Estimated sale price per unit in month 13 £100 000 $\times (1.05)^{13/12}$ = £105 428

(b) Costs

Services £45 000, £15 000 per month over three months.
Total building costs: £50 000 $\times$ 24 = £1 200 000
spread over eighteen months of construction = £66 667 per month of construction.
Building cost increasing at 8% per annum, that is 0.65% per month, so that the building cost in month 4 is £68 399.

Using these base data a cash flow analysis of the problem can be constructed as shown in Table 5.1.

Table 5.1 Residential development cash flow residual (£)

Period	Building	Architect/QS	Profit	Sale fees	Promotion	Outflow	Income	Net cash flow	Capital outstanding	interest
1	15 000					15 000		-15 000	-15 000	-150
2	15 000					15 000		-15 000	-30 150	-302
3	15 000					15 000		-15 000	-45 452	-455
4	68 399					68 399		-68 399	-114 305	-1143
5	68 839					68 839		-68 839	-184 287	-1843
6	69 282				10 000	79 282		-79 282	-265 412	-2654
7	69 728					69 728		-69 728	-337 794	-3378
8	70 176					70 176		-70 176	-411 348	-4113
9	70 628					70 628		-70 628	-486 090	-4861
10	71 082	75 000				146 082		-146 082	-637 033	-6370
11	71 540					71 540		-71 540	-714 943	-7149
12	72 000				10 000	82 000		-82 000	-804 093	-8041
13	72 463		31 628	6326		110 417	210 856	100 438	-711 695	-7117
14	72 929		31 757	6351		111 038	211 715	100 677	-618 136	-6181
15	73 399		31 887	6377		111 663	212 577	100 915	-523 402	-5234
16	73 871		32 016	6403		112 291	213 443	101 152	-427 484	-4275
17	74 346		32 147	6429		112 923	214 313	101 390	-330 369	-3304
18	74 825		32 278	6456	10 000	123 558	215 186	91 628	-242 044	-2420
19	75 306		32 409	6482		114 197	216 063	101 865	-142 599	-1426
20	75 791		32 541	6508		114 840	216 943	102 103	-41 923	-419
21	76 278	75 000	32 674	6535		190 487	217 827	27 340	-15 002	-150
22			32 807	6561		39 369	218 714	179 346	164 193	1642
23			32 941	6588		39 529	219 605	180 076	345 912	3459
24			33 075	6615	10 000	49 690	220 500	170 810	520 181	

Gross residual 520 181

Less acquisition costs at 3.00%

Net residual (maximum site bid: MSB) 397 745

Table 5.2 Variables in the residual (with current values)

Period of construction	18 months
Architects' fees	4.75%
Cost per unit now	£50 000
No. of units to be built	24
Thus total cost now	£1 200 000
Change in building costs	8.00%
Construction cost per month @ month 4	£68 399
Cost of borrowing per calendar month	1.00%
Current selling price	£100 000
Change in selling price p.m.	0.41%
Acquisition fees	3.00%
Agent's fees on sale	3.00%
Developer's profit	15.00%
Total investment period	24 months

The maximum site bid (MSB) is the present value of the residual discounted at 1% per month, less the acquisition costs. Overall, this tells the developer that if the values of the various factors in the model were to be those presented in Table 5.2, the maximum amount available to buy the land, the residual, would be just under £400 000.

Variability in the model

It is implicit in the analysis that there is variability present within the model. As a first step in examining the resulting variability of the residual, Table 5.2 gives the elements of the problem which might be regarded as variable, and the initial values used in the cash flow above.

Sensitivity analysis

The variability may be demonstrated to a certain extent by applying sensitivity analysis, aided by the computer model of the appraisal. For example, Table 5.3 shows the residual figures resulting when building cost and the monthly interest rate are varied. All the other variables are kept at their initial values. Note: since interest is charged monthly the equivalent annual rates are:

0.5% per month is 6.167%
1.0% per month is 12.683%
1.5% per month is 19.561%

This variability is seen in graphic form in Figure 5.1.

Figure 5.2 gives a further graphic description of this variability. This shows the effect on the residual value as developer's profit changes, at

Table 5.3 Sensitivity analysis: change in building cost/rate of borrowing

Cost of borrowing (%)	Change in building cost (%)					
	5	6	7	8	9	10
0.5	£512 510	£501 095	£489 656	£478 194	£466 707	£455 197
0.6	£495 135	£483 885	£472 611	£461 315	£449 995	£438 652
0.7	£478 172	£467 084	£455 973	£444 840	£433 683	£422 505
0.8	£461 611	£450 683	£439 732	£428 759	£417 764	£406 746
0.9	£445 445	£434 673	£423 879	£413 064	£402 227	£391 368
1.0	£429 662	£419 044	£408 405	£397 745	£387 063	£376 360
1.1	£414 254	£403 788	£393 301	£382 793	£372 264	£361 715
1.2	£399 213	£388 896	£378 558	£368 200	£357 822	£347 424
1.3	£384 531	£374 360	£364 169	£353 958	£343 728	£333 478
1.4	£370 197	£360 170	£350 124	£340 058	£329 973	£319 869
1.5	£356 206	£346 321	£336 416	£326 493	£316 551	£306 590

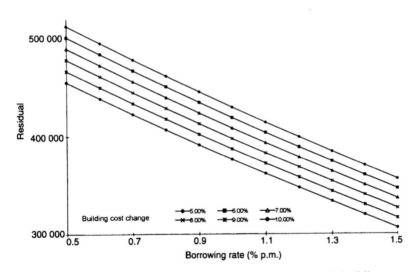

Figure 5.1 Sensitivity analysis: borrowing cost against change in building cost

different levels of building cost. At each level of building cost change, there is a constant reduction in developer's profit. Thus, with an 8% rise in building costs, each 1% increase in developer's profit leads to a reduction of £5333 in the residual funding available to buy the land. It should be noted that although Figures 5.1 and 5.2 appear to show a straight line

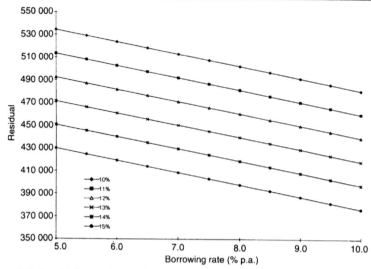

Figure 5.2 Sensitivity analysis: change in borrowing rate against developer's profit

relationship between these variables, the actual relationship is not quite linear, with the rate of change increasing as the percentage change in building cost rises.

If the residual's variability is to be examined by simulation methods, then a more detailed consideration has to be given to the nature of the variables enumerated in Table 5.2.

Identification of probabilistic variables

In this example, most of the variables can be regarded as:

1. directly controllable to a greater or lesser extent; others are
2. indirectly controlled, by calculation within the residual; a few are
3. uncontrolled and uncertain

Looking at each of these categories in more detail:

1. The number of units to be built (24) is controlled directly by the planning permission given, and could only be varied with great difficulty. This would also be true for the design and hence size of each of the units. The building cost is known, currently £50 000 per unit, as is the current sale price, £100 000 per unit. The developer's profit is fixed by the developer and is unlikely to be changed without a marked change in the decision-maker's attitude. This might arise if there were considerations other than pure profit in the appraisal.
2. Some variables are dependent upon others, fees being the principal

example, depending upon construction costs or sales income. They are regarded as controlled in this example.

3. (a) Variables which are to some extent uncontrolled are the period of construction, the number of units available at the start of month 13 when the units are first completed, and the total period of the investment. Careful internal organization, and the employment of a good building contractor could enable these variables to be kept sufficiently under control to make their values certain within reasonable limits.

   (b) Particular consideration has to be given to the following uncontrolled variables:
   (i) the change in building costs;
   (ii) the rate of borrowing;
   (iii) the inflation in sale price.

While variations in the other variables can be dealt with by adjusting their values singly within the simulation, as will be seen below, these three variables are the only elements in this example which would be better described by probability distributions. The present building cost is known. The rate and direction of change, even on average, is not known, but may be estimated. This comment is also true for sale price and any inflation in its value. Unless a fixed borrowing rate is negotiated in advance, changes in the cost of borrowing during construction may have a marked effect on the viability of the development.

These three variables can also be treated as effectively independent. It could be argued that the inflation of building costs and prices will be each related to overall inflation and are a part of it. Similarly changes in the borrowing rate will depend on the general state of the economy. The direct links between these variables are marginal, hence the view that these variables are independent. Later, a method for including some aspects of dependence will be examined.

The probabilities for these variables have been subjectively assessed, and this would be usual for an analysis of this kind, since the long-run (objective) data are likely to be limited, although in Chapter 6 an alternative approach for dealing with this aspect will be considered.

5.3 CASE STUDIES

Probability distributions

The distributions used to simulate the likely performance of these three variables are shown in Figure 5.3. It is assumed that these distributions have been constructed using the kinds of methods discussed in Chapter 3.

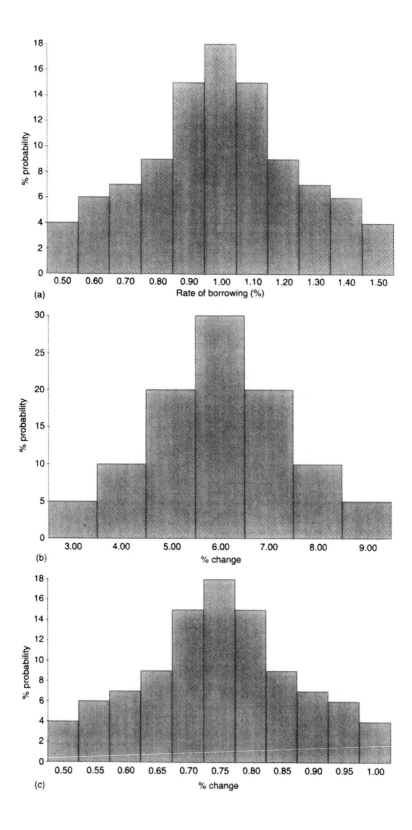

(a)

(b)

(c)

The values of all the other variables are determined, before the model is run on the computer.

These distributions have 11, 7 and 11 classes, and those in Figure 5.3(a) and (c) are distributed approximately normally (see Chapter 6 and the appendix).

These distributions are discrete, that is they cover a finite number of values. As such the variables can only take the values given. Thus, for example, the monthly rate of borrowing can be 0.8%, but not 0.827%. Sometimes a discrete distribution will model the variable accurately, but often it is an economical approximation of a continuous distribution. In a continuous distribution, the variable can, in theory, take any value along the scale between the limits defined for that variable.

Some general comments

While the user has to specify these three distributions, the underlying principle of simulation must be kept in mind. A theoretical population of all possible results for the MSB is being sampled, and the sample size is therefore of some significance. There are a number of factors to be taken into consideration when deciding sample sizes for simulation:

1. An analysis of this kind is relatively expensive. Unless a ready-made package program is available, cost may make possible only a small number of rather small sample runs. It may be that even if a package is available, results have to be to hand very quickly, and this will probably limit the size of sample obtained.
2. This expense has to be related also to the overall size and importance of the project. A large, complex development may need and repay considerable analysis, and larger sample sizes will provide more reliable estimates of the performance of the problem, which should in turn lead to more secure decisions.
3. Taking the three variable distributions shown in Figure 5.3 their ranges of values can generate 847 (11 × 11 × 7) possible combinations. This total will of course change according to the number of classes in a discrete distribution, and will increase if any other random variables are added to the analysis, or if continuous distributions are used. If the sample is small it will be difficult to cover even a proportion of these possible combinations.
4. Related to these combinations are their probabilities of occurrence. In this case, the least probable combination may be expected to occur

---

Figure 5.3 Distribution of (a) possible monthly borrowing rate; (b) change in building costs; (c) change in house price per month

once in 10 000 results. Other things being equal, if the performance of the development is to be modelled satisfactorily, the number of samples taken should be large enough to allow a reasonable number of these low probability events to be modelled. Because other things are not usually equal, a trade off has to be made against these low probability events, and much smaller sample sizes are common, indeed these large sample sizes are rare in business-type simulation models.

5. On the other hand sample sizes should not be too small. A single result is not appropriate in simulation, because the method relies on sampling a statistically valid number of times. A rule of thumb minimum number of sample results is 30. In any case a number of different sized samples should be taken to determine the stability of the distribution of results.

To carry out the simulation it is necessary to

1. sample the values for the state variables;
2. insert them into the appraisal model along with the determined values;
3. make the cash flow calculations to find the MSB;
4. and repeat until the required number of observations has been obtained.

Within a spreadsheet this may be done by means of a macro, which is a set of recorded and stored calculation steps, and the macro can be repeated the desired number of times (see for example Mollart, 1988).

In EXCEL version 5, however, macros, although still usable, have been effectively replaced by 'modules'. These are mini computer programs written in a language called Visual Basic, which is built into the spreadsheet and is always available. The use of Visual Basic modules, together with the standard functions of the spreadsheet, allows very sophisticated models to be constructed in the spreadsheet, and assists greatly in the process of Monte Carlo simulation. Here therefore, the simulation models are 'driven' by specially prepared Visual Basic modules that control the Monte Carlo sampling process, insert the sampled values into the cash flow model, and assemble the results for samples of various sizes.

For illustrative purposes, the results from the simulation have been broken down into a series of separate sections, demonstrating the various features of such an analysis.

Two state variables: an initial analysis

Although three variables have been identified that can be reasonably represented by probability distributions in this model, in the first instance only two of these will be included in the simulation. These are the monthly borrowing rate and the change in building cost. For the moment the change in house price variable remains fixed at its original value.

Each complete set of results will be described as a 'run'; any single result is one 'cycle' or iteration within any run. It would, of course, be perfectly possible to make the computer display or print out all the intermediate cash flow figures for each cycle. In this case, this would mean displaying the contents of Table 5.1 for every cycle. This would only be sensible for a very limited simulation exercise, since the detail, and amount of paper, produced would rapidly swamp the decision-maker, to very little useful purpose.

Only one probability distribution is used for each of the state variables. For short time scale projects it would be uneconomic to produce a series of time-based distributions for each variable. Here the time base is too short and therefore the distributions used should be regarded as averages of the continuous series of distributions which theoretically exist throughout the life of the project. Computationally there is little difficulty in accommodating such extensions; the computer would be instructed to select the appropriate distribution for each variable depending upon the time period being considered, rather in the manner suggested in Chapter 4.

To demonstrate the results obtained, Tables 5.4(a) and (b) show the output from a run of 100 cycles. Each value is the final maximum site bid. In Table 5.4(a) the results are presented in the order in which they

Table 5.4(a) Maximum site bids from 100 cycle simulation (results as drawn)

370 197	456 195	380 205	424 700	408 405	408 405	466 923	429 662	397 745	360 170
435 125	378 558	419 044	366 073	456 195	402 227	495 135	440 258	419 044	435 125
402 227	403 788	456 195	402 227	397 745	388 896	375 920	403 788	388 896	434 673
419 788	423 879	414 254	357 822	500 279	434 673	445 445	423 879	512 510	336 416
393 301	428 759	336 416	423 879	340 058	439 732	414 254	450 683	408 405	382 793
417 764	378 558	403 788	336 416	424 700	434 673	370 197	419 044	408 405	440 258
517 565	382 793	336 416	467 084	461 315	489 237	461 611	429 662	366 073	429 662
434 673	444 840	450 683	350 124	353 958	399 213	466 923	434 673	414 254	423 879
419 044	434 673	506 362	399 213	366 073	429 662	408 405	414 254	461 611	461 315
455 973	419 044	461 611	483 402	419 044	393 301	483 402	450 683	378 558	461 611

Table 5.4(b) Maximum site bids from 100 cycle simulation (results in rank order)

336 416	366 073	382 793	402 227	414 254	419 044	429 662	435 125	456 195	466 923
336 416	366 073	388 896	402 227	414 254	419 788	429 662	439 732	456 195	467 084
336 416	370 197	388 896	403 788	414 254	423 879	429 662	440 258	456 195	483 402
336 416	370 197	393 301	403 788	414 254	423 879	434 673	440 258	461 315	483 402
340 058	375 920	393 301	403 788	417 764	423 879	434 673	444 840	461 315	489 237
350 124	378 558	397 745	408 405	419 044	423 879	434 673	445 445	461 611	495 135
353 958	378 558	397 745	408 405	419 044	424 700	434 673	450 683	461 611	500 279
357 822	378 558	399 213	408 405	419 044	424 700	434 673	450 683	461 611	506 362
360 170	380 205	399 213	408 405	419 044	428 759	434 673	450 683	561 611	512 510
366 073	382 793	402 227	408 405	419 044	429 662	435 125	455 973	466 923	517 565

were sampled by the computer. In Table 5.4(b) they have been rearranged in rank order. It will be seen from Table 5.4(b) that a quite considerable number of results occur more than once. This is to be expected because of the way in which the probabilistic variables are likely to combine.

Clearly, in this form, any interpretation is very difficult. For this reason an astute first step is to prepare a histogram, classifying the results into a rather more sensible form. This is shown, using the Table 5.4 figures, in Table 5.5, and in Figure 5.4, as both a histogram and a cumulative frequency.

The expected value $E(X)$ and the standard deviation (s) of the distribution are also calculated:

Table 5.5 Frequency distribution for 100 cycles

Less than £s	Frequency	Cumulative %
315 000	0	0.0
340 000	4	4.0
365 000	14	23.0
415 000	21	44.0
440 000	28	72.0
465 000	17	89.0
490 000	6	95.0
515 000	4	99.0
540 000	1	110.0

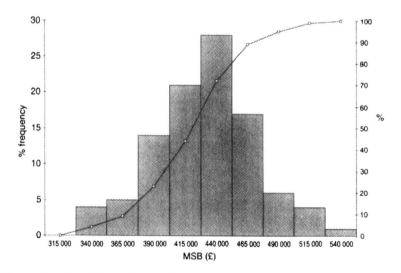

Figure 5.4 Simulation results, 100 cycles: histogram and cumulative distribution

$$E(X) = £419\,466.5, \quad s = £40\,718.6$$

The 'shape' of this distribution is affected by two factors:

1. The 'shapes' of the distributions used to represent the state variables.
2. The monthly rate of interest variable is distributed symmetrically around the value used in the original example, i.e. 1% per month rate of interest, although the probability attached to that value is quite low (0.18). The change in building cost variable is distributed around a slightly different value to that originally used (0.65% per month), but with a fairly high probability of 0.30.

Because of this it may be expected that distributions of sample runs will show highest frequencies for values close to that originally determined and will be 'spread' to reflect the combinations of values which may arise from the state variables used.

Among other things, the frequency distribution shows that the influence of these probability distributions is to give 29 chances in 100 that the MSB amount will be less than £400 000, say £397 745, the original determined value.

This latter value has relatively little real impact in the simulation. The probability distributions used are the factors that affect the values which are output from the simulation and it is these that now make such an individual result only one of many possible, only 100 of which have so far been selected at random.

The results from this run can now be tested for stability before being interpreted.

Repetition, stability and variability: testing the model

The real advantage of computerized models is the ability to repeat. Given the general comments made earlier, repeat runs may therefore be made as many times as is expedient, but using different series of random numbers to generate the values for each run.

The results of a series of different runs are shown in Figure 5.5. This depicts the standard form of output for a simulation, a cumulative relative frequency distribution (CRF). A histogram may also be included but the CRF is perhaps interpretatively more useful.

In Figure 5.5, ten runs of 100 cycles each are plotted. The statistics for each run, the expected value, $E(X)$ and the standard deviation, are shown in Table 5.6.

The statistics in Table 5.6 seem, even without reference to Figure 5.5, to suggest that the distribution of residual values from the variables is quite stable, although as the diagram shows there are differences in the detailed shapes of each distribution. The difference between the highest and lowest mean values in these ten runs is less than 5%.

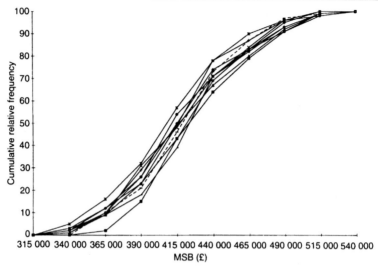

Figure 5.5 Cumulative relative frequencies: ten runs, each 100 cycles

Table 5.6 Statistics for ten runs in Figure 5.5

Run	E(X)	SD
1	417 362.7	42 925.2
2	429 584.6	38 099.0
3	423 798.0	42 542.9
4	418 563.7	42 507.7
5	409 660.6	41 863.8
6	418 796.3	42 725.8
7	416 411.8	39 690.5
8	424 412.3	40 820.3
9	419 621.3	37 925.4
10	423 000.4	44 757.8

Reducing the sample size to 50 cycles per run, Figure 5.6 shows the CRFs for ten runs of this size. A sample size of 50 is relatively small, and might be expected to show greater variability, because of the smaller sample size, than the samples of size 100.

The CRF shapes are indeed less compact than in Figure 5.6. The statistics for these runs are shown in Table 5.7. These confirm the greater variability. Although the difference between highest and lowest mean is now closer to 2%, the standard deviations are much less similar.

This straightforward comparison of these two groups of runs serves to demonstrate clearly the decrease in variability which can be achieved by increasing sample size. These runs exhibit variability due to sampling.

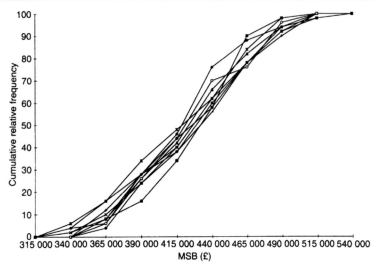

Figure 5.6 Cumulative relative frequencies: ten runs, each of 50 cycles

Table 5.7 Statistics for ten runs in Figure 5.6

Run	$E(X)$	SD
1	419 151.9	39 969.2
2	425 418.4	35 074.2
3	423 373.4	41 776.8
4	426 164.0	45 922.0
5	415 671.4	46 925.5
6	428 019.7	42 060.0
7	425 465.4	46 084.1
8	426 958.4	45 454.4
9	424 508.4	39 009.7
10	424 034.6	48 628.3

Overall, however, the results are still of the same order of magnitude, and suggest that the values observed are sufficiently stable to be usable for decision-making.

To show the effects of a much increased sample size, Figure 5.7 shows a histogram and CRF for a run of 2000 cycles. The mean of this distribution is £419 107.9. The standard deviation is 41 196.3. It has to be remembered that variability will never be completely removed at any stage: it is inherent in both the sampling process and in the use of probability distributions for some variables in the analysis. The standard deviation, measuring variability and thus expressing uncertainty, is, however, noticeably reduced by comparison with earlier runs.

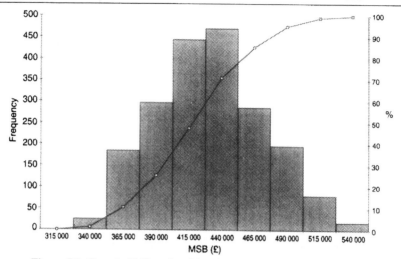

Figure 5.7 Case 1, 2000 cycles: histogram and cumulative frequency

Some other subtle differences can be seen. In particular that by far the largest proportion of observations are in the class from £460 000 to £490 000. The $E(X)$ appears relatively low by comparison with most seen earlier, and the standard deviation is very much in the middle of the range seen in Table 5.6.

Interpreting the model's performance

How are these outcomes to be interpreted in a decision context?

Thus far, the variability of runs has been discussed by reference to histograms and cumulative relative frequency distributions. These are directly comparable and are independent of varying sample size.

Twenty-two separate runs of differing sizes have been described and compared in this way, simulating the occurrence of a total of 3600 values of the MSB. As long as the basic variable structure remains fixed, then the results of individual runs can be combined, because each result is effectively independent of any other. Figure 5.8 shows the histogram and cumulative distribution for all 3600 cycles run on these variables. In this distribution the relative frequencies are expressed, not as proportions of all observations, but more explicitly as if they are the long-run probabilities of occurrence for any value of the new random variable MSB which has been generated by the simulation. This cumulative distribution is formally known as a cumulative density function (CDF), indicated primarily by the use of probabilities on the vertical axis of the graph.

The overall mean of this sample is £420 061.6 with a standard deviation of 41 560.6. This mean value is in effect the mean of a set of means. A

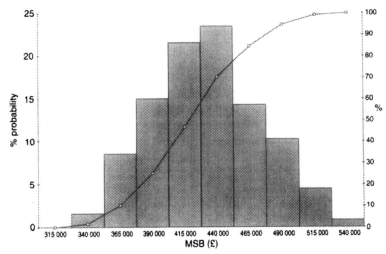

Figure 5.8 Combination of runs, 3600 cycles: histogram and cumulative frequency (CDF)

run of 3600 cycles is a large 'sample'. It is greater than the actual number of alternative combinations in this case. This being so it may be regarded as reasonably representative of the population characteristics of the MSB and can be said to model the long run performance of the problem.

This large sample distribution will be used for further analysis.

Measures of central tendency: the average performance of the model

The first step in the analysis of these distributions is a consideration of the measures of central tendency, the mean, median and mode. This can be done directly within the spreadsheet, using the Excel Descriptive Statistics analysis tool. The useful values generated by this command are given in Table 5.8.

The minimum value observed in this set of 3600 observations is well below the original determined value, although the effect of the probability

Table 5.8 Statistics of 3600 cycles

Mean	420 061.6
Standard deviation	415 60.6
Coefficient of variation	0.099
Median	419 043.9
Mode	419 043.9
Range	218 717.6
Minimum	316 551.1
Maximum	535 268.6

distributions gives that minimum value a very low probability of occurring! At the other end of the range, there is a small chance that as much as £535 268 could be available as the MSB, and again this has a rather low probability of happening.

The mean, as the expected value, represents the long-run average performance of the MSB under all conditions. All the conditions which can arise are all the possible combinations of variables in the model.

The standard deviation measures the spread of the distribution and as discussed earlier, provides the basic measure of the 'risk' in the system. The coefficient of variation is also given here to provide a comparison of the variety of distributions which are given in this chapter. Large differences in this coefficient would merit special attention in the analysis.

Generally the mode or modal value is the value of any variable that occurs most frequently. It may not be possible to determine the modal value directly, although there is the possibility that all values might be inspected, to discover which occurs most often. (It can also be found by calculation, using methods found in most elementary statistics texts.)

Expressed in probability terms, the mode is the most likely value, and as such has some significance in that it is the value which will have the most positive instant influence on the decision-maker. It could be argued that if it is the most likely value then should it not be given extra weight when the results of the simulation are to be used for a decision? This is sometimes true, but the mode should not be considered in isolation. Looking at the histogram in Figure 5.8, a modal class from £415 000–£440 000 can be seen. Any value in the modal class in Figure 5.8 has a 0.25 probability of selection. One class lower, any value has a 0.22 probability, a difference which might be felt to be rather marginal. Taken together, however, these two classes include nearly half of the distribution in a range of £50 000. This information might be more valuable than a much narrower view of the mode itself (this value appears 216 times in the series, and has a probability therefore of 0.06). This will be particularly so in cases like the present one, where the modal class does not have a significantly higher probability than others in the distribution as a whole. There may well also be cases where distributions are bimodal (having two classes of equal 'greatest' frequency) or multi-modal (having a number of classes of equal greatest frequency).

The median value lies at the 50:50 point in the cumulative distribution. Thus, the median lies at the equal probability point of the distribution. If unlike here, the distribution cannot be tabulated and the median value found by inspection, then its position can be measured geometrically as is done in Figure 5.8. The computed value here is marginally less than the mean at £419 044, and is, coincidentally, the same as the mode!

Using this geometric approach, the probability of the residual being less than or more than any value can be assessed, as can the probability of occurrence between any values. As an example, suppose that this land is actually on offer at a known price, say £500 000. This then becomes in effect a further, external, uncontrolled but determined variable in the model.

Given the case 1 variables, what is the probability that this amount will *not* be available? That is, what is the chance that the variables will combine to produce an MSB less than £500 000, making the development non-viable?

Looking at the CDF in Figure 5.8 and reading off the cumulative probability curve for a residual value of £500 000, the cumulative probability is 0.96. That is, 96% of observed values lie to the left of £500 000 and are less than this value. There is therefore just slightly less than a 1 in 20 chance that there will be as much as £500 000 available to finance the land purchase (1.00–0.96). Put the other way, if the land was purchased at this price, the development would have only a 0.04 probability of making at least the required profit. Conventionally this will be allowed for to a satisfactory extent by the contingency element in the 'profit' calculation. While a real loss may not be incurred, it will be for the decision-maker to assess whether this is an 'acceptable' reduction in profit and return. The model may be used to assist in this process.

Sensitivity analysis in simulation

If the developer is unhappy with the level of risk which is implied by the results of the analysis shown above, consideration might be given to changing the value of any variable(s) which are controlled by the decision-maker. The sensitivity of the simulation output to such changes can be examined in exactly the same way as in any other sensitivity analysis. The value of the variable is simply replaced in the spreadsheet. Which variable(s) could the developer consider? One which is clearly in his control is the level of desired profit. What would be the possible consequences of varying that profit? Clearly the loss probability in Figure 5.8 can be reduced by decreasing the level of profit required, and the extent of this can then be reviewed. At the same time the effect of increasing the profit percentage might be looked at as well.

(a) Developer's profit

Figure 5.9 shows CRFs for simulations using the two state variables: but increasing developer's profit to 20%, and then decreasing developer's profit to 10%. The central CRF is for the 15% developer's profit from Figure 5.4.

The run size is 100 cycles in each case, and the relative variability of these smaller sample sizes may be compared with the smooth CDF of the

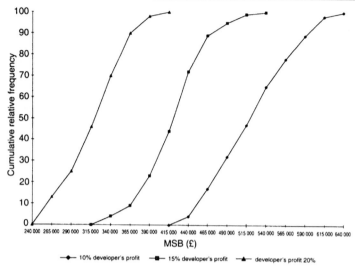

Figure 5.9 Sensitivity analysis on developer's profit, 100 cycles

very large sample size in Figure 5.8. Also, placing these three rather differently shaped CRFs side by side gives a good impression of between-sample variability. Even so, the limited run size is felt to be sufficient on the basis of the results from earlier tests.

Comparatively the figures in Table 5.9 are extremes. With developer's profit set to 20% there is no possibility of a positive MSB if land price is £500 000, since all observed results are less than £390 000. By contrast, at 10%, there appears to be a probability of about 70% that enough residual value will be available to fund the land cost, since the lowest possible residual is greater than £415 000.

The model is flexible enough for a whole series of CRFs to be constructed given different levels of developer's profit, so that if desired the decision-maker could assess the probability, and therefore estimate the

Table 5.9 Run statistics changing developer's profit (£)

	Developer's profit (%)		
	10	15	20
Mean	520 926.0	419 466.5	317 429.2
Standard deviation	50 999.0	40 718.6	38 313.4
Coefficient of variation	0.098	0.097	0.121
Median	523 553.2	419 043.9	318 230.5
Mode	523 553.2	434 672.6	314 534.6

risk of not making a required profit for a given land price. The alternative might be to attempt to renegotiate the land price – downwards!

(b) Building cost

Variables which are only capable of partial control may also have their sensitivity tested in this way. As an example of this we can consider building cost. If the cost of constructing each unit could be reduced, say by the use of different materials, then the MSB would be increased. How much cost per unit would have to go before there was a zero probability that £500 000 would not be available, given that we do not know what the change in building cost will be, after the development has started?

Figure 5.10 shows the CRFs for four samples of size 100, where building cost has been set at differing values as shown in Table 5.10. (The curve for unit building cost = £50 000 is the same as that for 15% developer's profit in Figure 5.9.)

The differences in the shape of these curves are again a function of the small sample size. If the sample sizes were sufficiently large, the CRFs would be of the same general shape, and be an equal distance apart for a unit change in building cost. They would then be equivalent to the deterministic sensitivity analysis shown in Figure 5.1.

It will be seen that in this case the probability of having more than £500 000 available, at a building cost of £50 000, is 0.025. This is marginally less than the probability of 0.04 derived above from the sample of 3600. The difference is again attributable to varying sample sizes.

As pointed out earlier this variable could, to some extent, be in the

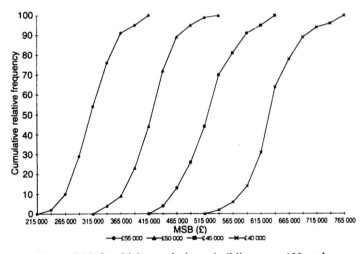

Figure 5.10 Sensitivity analysis on building cost, 100 cycles

Table 5.10 Run statistics for various unit building costs

	Building cost (£)		
	£40 000	£45 000	£55 000
Mean	633 978.29	521 493.25	312 458.59
Standard deviation	46 217.29	48 773.21	39 435.68
Coefficient of variation	0.073	0.094	0.126

developer's control. From Figure 5.10 it is clear that building cost has to be considerably less than £45 000 before there is a zero probability that the MSB amount will take a value less than £500 000, although if the building cost can be reduced to £40 000, then this clearly will be the case. The decision-maker has two problems:

1. Can costs be reduced to £40 000 from a known £50 000? If not, can costs be reduced at all?
2. If costs can be reduced only to £45 000, is the 50 + % chance of cutting into profit acceptable? Is any chance of profit reduction acceptable?

If not, the decision-maker must accept that the project is not viable.

(c) The rate of borrowing

As a final example simulating the MSB's performance using these two distributions, consider the possibility that funding might be negotiated at a fixed interest rate. The MSB's sensitivity can then tested at different annual rates of borrowing.

The crucial point here is that since the cost of borrowing, although sensitive, is effectively certain, a significant amount of uncertainty is now removed from the simulation. All the other variables in the model keep the values initially assigned to them and thus only one state variable (change in building costs) remains in the model.

The effect of this reduction in uncertainty can be seen in the histogram in Figure 5.11. The range of MSBs is substantially reduced, and there is a marked modal class, containing almost 60% of all observations. To test the sensitivity of this model to changes in annual borrowing rate, the results of three simulation runs are shown in Figure 5.12 and are statistically summarized in Table 5.11. The values selected for the analysis are the most likely, 1% per month, 12.68% per annum, the most pessimistic (highest), 1.5% p.m.; 19.56% p.a., and the most optimistic (lowest), 0.5% p.m.; 6.17% p.a. taken from the probability distribution for the variable (Figure 5.3a).

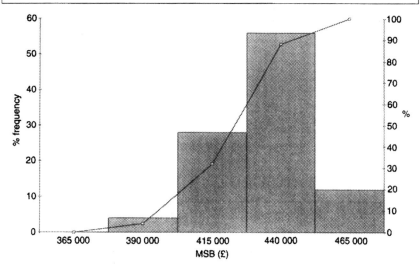

Figure 5.11 One state variable, borrowing rate: histogram and cumulative frequency over 100 cycles

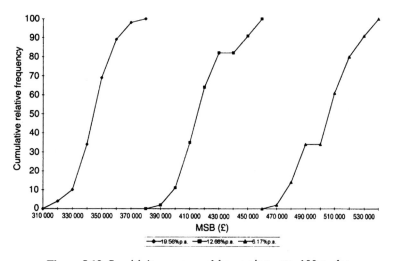

Figure 5.12 Sensitivity on annual borrowing rate, 100 cycles

The CDFs are 'compressed' by comparison with all those discussed earlier. This is confirmed by the substantially reduced standard deviations. Even allowing for the relatively small sample sizes, the decreased uncertainty in the model is reflected in consistently reduced variability, as expressed by the standard deviations, and in a tendency to bunch in the sample results clearly seen in Figure 5.12.

Variability in the overall performance of the MSB is not markedly

Table 5.11 Run statistics for various borrowing rates

	Monthly rate (%)		
	0.5	1.0	1.5
Mean	503 124.6	420 615.6	345 909.9
Standard deviation	17 282.9	15 496.3	12 496.1
Coefficient of variation	0.034	0.037	0.036

affected by changes in the annual borrowing rate, certainly by comparison with the magnitude of changes observed in the results when the building costs were considered earlier.

More significantly, however, fixing the borrowing rate does not appear to help greatly if the land costs £500 000. Indeed, with interest fixed at 12.68% per annum (1.0% per month), there appears to be no chance that there will be at least £500 000 available to purchase the land. At 6.17% per annum (0.5% per month) that probability rises to about 36%, meaning that about two-thirds of the time the purchase price threshold would not be reached.

The rather higher probability of achieving an MSB of £500 000 discussed earlier was obtained with more uncertainty in the system. It could therefore be described now as misleading, even though the sample size was very large. The additional uncertainty, built into the model by using the distribution of the random variable 'rate of borrowing' rather than the single values now being used, produced a rather more optimistic view of the probable outcome of the MSB. This illustrates one of the paradoxes of this kind of analysis, that the availability of more information, though a reduction in uncertainty can actually produce results which seem 'worse' than those obtained in the absence of that information! The problem is that this additional information is not often available and we may have to make do with the 'optimistic' probability-based view.

Three state variables

Having seen the reduction in uncertainty which results when one of the two probabilistic variables in the model is made determined, the next step is to examine the effect of adding more uncertainty in the form of the third of the state variables identified earlier, the percentage change in unit selling price per month (Figure 5.3(c)).

Because higher completion prices are now possible, with a price inflation of up to 1% per month, the frequency distribution is extended at higher values. It is also shifted up the scale because the possible values of

Table 5.12 Three state variables: frequency and cumulative frequency

Less than £'s	Frequency	Cumulative %
390 000	1	1.0
415 000	1	2.0
440 000	7	9.0
465 000	6	15.0
490 000	16	31.0
515 000	9	40.0
540 000	15	55.0
565 000	15	70.0
590 000	8	78.0
615 000	12	90.0
640 000	5	95.0
665 000	2	97.0
690 000	3	100.0

Table 5.13 Run statistics for 100 cycles, using three state variables

Mean	531 941.7
Standard deviation	64 658.3
Coefficient of variation	0.122

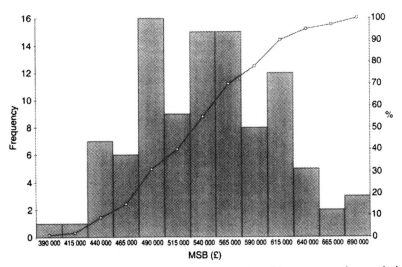

Figure 5.13 Three stochastic variables, 100 cycles: histogram and cumulative frequency

the state variable start at a higher value (0.5% p.m.) than that originally used (0.41% p.m.).

The implication of this is that the expert responsible for assessing the probability distribution of this variable has reconsidered its likely performance since it was first assessed, and as a consequence appears to have increased the probability of achieving MSBs close to or above the purchase threshold of £500 000. As Figure 5.13 shows, there is now a probability of reaching £500 000 or more of about 62%, a much larger value than any observed thus far.

Figure 5.13 also shows the effects of uncertainty and small sample sizes. The frequencies in the classes are highly variable, and the pattern of the cumulative frequency curve is erratic. The standard deviation of the run is increased, as is the coefficient of variation, and the range of values is 308 000, compared with 181 000 for the initial example (Figure 5.4).

This particular set of variables (shown in Figure 5.3) will be used again in Chapter 6 to explore further elements of the simulation process.

Changing the shape of a distribution

In the next example, all three state variables remain in the model. However, the rate of borrowing variable now has a changed probability distribution, that shown in Figure 5.14.

It will be seen that in this case the expert financial analyst feels that interest rates are very likely to rise above the current level of 1% per month. It is also believed that there is only a very small chance (0.05) that

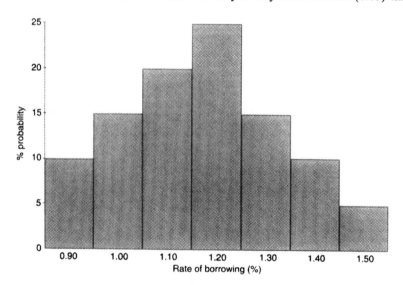

Figure 5.14 New distribution of possible monthly borrowing rate

Table 5.14 Run statistics, interest rate
variable distribution changed

Mean	506 519.6
Standard deviation	48 478.9
Coefficient of variation	0.096

the interest rate will fall below 1% per month, and that it will not fall
below 0.9% per month. Notice also that the distribution has fewer classes
(seven) than before and that it is no longer symmetrical.

The overall results of this change can be seen in Table 5.14 and Figure
5.15. To counter the variation due to small sample size, this has been
increased to 300 in this case, and the result is a much better shaped histogram.

Having observed the effect of previous changes on the results of the
simulation, it is possible to see the combined effects of having three state
variables in the model, of changing the shape of the distribution and of
increasing the sample size.

1. Because the probability distribution now being used to model the
   performance of the rate of interest is truncated relative to that shown
   in Figure 5.3(a) any variability due to this source should be reduced.
   The overall range for this distribution is approximately 293 000, less
   than the previous case, again caused by the new distribution. The

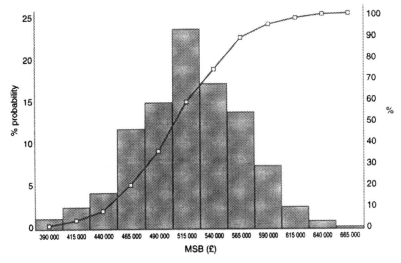

Figure 5.15 Three state variables, revised distribution of price change: histogram
and cumulative frequency

results shown in Table 5.14 should be compared therefore with the results from Table 5.13 where the variability is much larger and with Table 5.11 where variability was eliminated for the variable 'rate of interest', causing the overall variability, as expressed by the standard deviations, to fall to below £20 000.

2. The run size in this case is 300, rather larger than the majority of previous samples used for analysis, which were mostly 100 cycles. Naturally an increased sample size will be expected to reduce the variability to some extent. It can be noted that increasing sample size will only decrease the amount of variation in proportion to the square root of sample size. Thus:

Sample size 100 $\sqrt{100} = 10$
Sample size 300 $\sqrt{300} = 17.32$
Difference   200   7.32

As a result of these factors, the mean value is reduced by just under 5%, although clearly there is still a reasonable probability of achieving a MSB of at least £500 000 (about 45% from Figure 5.15).

Developer's profit at 10% and 20%

As a final illustration of the way in which the simulation may be tested, developer's profit has been varied and two runs have been made, with developer's profit at 10% and 20% as well as the original 15% (Table 5.15). The CDFs are shown in Figure 5.16.

There are clearly substantial differences between the results of these runs and those in Table 5.10. Now a reduction in developer's profit to 10% increases the amount available by nearly £100 000, and increasing the profit to 20% still increases the amount by more than £70 000. It would be expected that these positive and negative differences would be of essentially the same order in this case were it not for the changed form of the distribution of the interest rate variable.

Table 5.15 Run statistics, changing developer's profit percentage

| | Developer's profit (%) | | |
	10	15	20
Mean	616 320.9	506 519.6	388 849.4
Standard deviation	54 296.9	48 478.9	44 024.0
Coefficient of variation	0.088	0.096	0.113

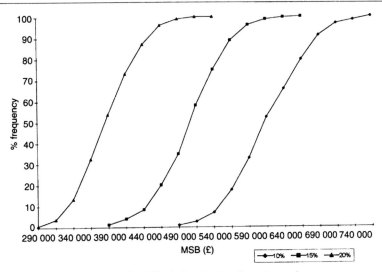

Figure 5.16 Variation in developer's profit

5.4 CASE STUDY CONCLUSIONS

What conclusions may be drawn from this analysis? In this example, changes in the shape of the distributions given (subjectively) to the state variables are seen to have relatively little effect on the amount available to purchase the land.

Changes to some other variables certainly affect the residual amount in a far more sensitive way. A somewhat more extensive sensitivity analysis would have to be performed to test which of the determined variables actually have the greatest effect. We have demonstrated that the profit which the developer demands has a marked effect in this example on the final result, of the order of +20% for a +5% change in the variable.

Given the range of values of the variables, all the residual values observed are substantially above zero, but as noted above the zero value may not be at all important. Much more significant in this analysis will be the price which is being asked for the land and the probability of achieving that result.

The simulation model has explicitly shown the interaction of these two variable types. It has demonstrated that in this example at least, the effects of uncertainty are limited, and in a sense predictable, since there was little change in results when the subjectively derived shapes of the probabilistic variables were changed. Much more variability, measured by the standard deviation of the distribution of results, was shown in this case when certain deterministic variables were adjusted. There is no reason to suppose that the most important variables in a simulation need be probabilistic.

In this chapter spreadsheets prepared in Microsoft EXCEL have been used to carry out a Monte Carlo simulation on a cash flow development appraisal. To do this it has been necessary to write macros or 'modules' in Visual Basic. On the whole this is a reasonably straightforward exercise, although it does require a certain 'programming' skill. There are now, however, very sophisticated ways of carrying out risk analysis that do not involve such programming, and which can add still more to the range of analysis which is possible. This is the subject of the next chapter.

Advances in development appraisal simulation | 6

6.1 INTRODUCTION

Although some of the methodology described was complex, Chapter 5 really looked at the basics of the simulation process. In particular it looked at how a decision-maker can interpret the results from the Monte Carlo sampling process. This chapter builds on that analysis and considers some of the more advanced elements of simulation that are now possible. A somewhat different approach is taken to do this.

As was demonstrated, simulations can now be written quite easily into spreadsheets. The simulations were carried out using Visual Basic modules written within the EXCEL spreadsheet. Even this is no longer absolutely necessary however. A spreadsheet 'add-in' is used in this chapter. An 'add-in' is an extra package or tool that, as the name implies, can be added to the spreadsheet. It links itself to the spreadsheet to carry out a specific range of tasks – in this case risk analysis. The availability of an add-in to do risk analysis is an indication of the relative significance and perceived value of the approach, in that, if a significant market for the product did not exist then these (professionally produced) add-ins would not have appeared. There are a number of such packages available, the one employed here is called @RISK* (see Mollart, 1994).

The major attraction of such a tool is the possibility of greatly increased flexibility in modelling the risk analysis process. It removes the need for extensive 'programming' or reprogramming of the spreadsheet if the basic appraisal model has to be substantially changed from one development to the next.

These products vary in complexity, but, they are capable of generating sophisticated output, with more detail than that demonstrated in Chapter 5. The complete @RISK package consists of several elements that can be used separately or as a comprehensive and systematic unit for risk analysis. There is, however, a price to pay because the methodology of the software system needs to be properly understood and appreciated, especially as the approach required can be rather different.

In this chapter, @RISK is used to explore a number of features of simulation that can be accomplished relatively simply by using this package. These are, interdependence between the state variables, the use of continuous probability distributions, and related to this the data requirements of such distributions. There will be less emphasis on the decision-making aspects and more on the technical elements of the simulation process and the way that they can be improved by the features offered in a package such as @RISK. This does not mean that the decision process can be ignored however, because judgements still have to be made about the shape and form that a simulation is to take. A system such as @RISK should not be used as a 'black box', with an uncritical view of the inputs and no awareness of the methods employed by the software system to reach the output which is presented to the user. With this aim in mind, the first section compares the approach taken so far with that required for @RISK.

6.2 USING @RISK

The development appraisal from Chapter 5 is used again here to show how a simulation can be carried out using a risk analysis add-in such as @RISK.

Using this model, a sample of 500 has been taken as a benchmark against which to compare the output from the @RISK system. The particular form of the model is that used to generate a run of 100 cycles in Chapter 5. This is shown in Tables 5.12 and 5.13 and Figure 5.13. Table 6.1 and Figure 6.1 show the new results.

How is this simulation approached using @RISK? Initially, there is little difference. The basic cash flow model, as defined in Tables 5.1 and 5.2, is set up exactly as before as a standard EXCEL worksheet.

It is of course necessary to distinguish the stochastic variables in the model. There are three state variables providing input values to this model: change in building cost per annum; change in house price per

Table 6.1 Run statistics for 500 cycles, using three state variables

Minimum	372 114.10
Maximum	728 839.17
Mean	531 365.99
Standard deviation	60 699.85
Coefficient of variation	0.114
Median	530 343.67
Mode	547 965.09

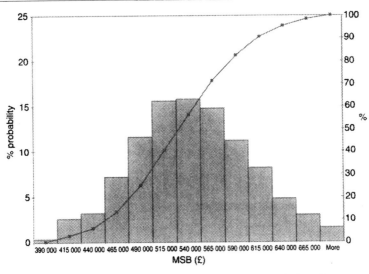

Figure 6.1 Three state variables: histogram and cumulative frequency, 500 iterations

month, and monthly borrowing rate. Their probability distributions were drawn up from the subjective views of expert advisers as best estimates of the expected performance for the variables, and their distributions are given in Figures 5.3(a), (b) and (c). Values for these variables are sampled using the Monte Carlo method. The Visual Basic module takes the distributions and draws a value randomly from each. The @RISK system does much the same thing, but each of the variables has to be defined in a somewhat different way.

Thus far these variables have been described as 'discrete'. The possible outcomes are limited to certain specified values, given in Table 6.2. They were presented in this way partly because a subjective probability assessment tends to produce such an approximating distribution, and also because programming a discrete distribution is quite easy to do in Visual Basic! The same distributions can be modelled using @RISK.

In this system, state variables are called input variables, and a 'description' has to be given for each of them. This is a description of the shape of each of the distributions. Instead of using the cumulative frequency to describe the distributions, @RISK uses the actual class frequencies or probabilities and these are shown in Table 6.2.

In the original spreadsheet model, at each cycle of the simulation, the sampled value of each of these variables was placed in its own specific cell. This cell was then used as the base for all further calculations using this value. Using @RISK, each of these same cells must now contain the description of the relevant distribution.

Table 6.2 Probability distributions for the three state variables

Variable	Cumulative probability	Probability
Change in building cost per annum (%)		
3.00	5	5
4.00	15	10
5.00	35	20
6.00	65	30
7.00	85	20
8.00	95	10
9.00	100	5
Borrowing rate per month (%)		
0.50	4	4
0.60	10	6
0.70	17	7
0.80	26	9
0.90	41	15
1.00	59	18
1.10	74	15
1.20	83	9
1.30	90	7
1.40	96	6
1.50	100	4
Change in house price per month (%)		
0.50	4	4
0.55	10	6
0.60	17	7
0.65	26	9
0.70	41	15
0.75	59	18
0.80	74	15
0.85	83	9
0.90	90	7
0.95	96	6
1.00	100	4

Suppose that we have to describe the variable 'change in building cost'. Assume that the seven values of the variable are in the spreadsheet cells G4 to G10, and that the probability values are in the cells I4 to I10. The sampled value of the variable is to be placed in cell D7. Thus in D7 a special @RISK formula is written in the form:

RiskDiscrete (G4 : G10, I4 : I10)

The same process is carried out for each of the other state variables using the appropriate cells.

The variable to be simulated is still the maximum site bid (MSB). As an output variable, its spreadsheet location has to be identified to the system.

Having established the input and output variables, the form that the simulation is to take has to be defined. This will be discussed in some detail since there are several important features which @RISK offers that can affect the way that the simulation proceeds, and these need to be set before starting the simulation.

The simulation settings menu from @RISK is shown in Figure 6.2. The number of iterations (cycles) is set to 500, to match those carried out for Table 6.1. This number of iterations is used in all the simulation runs carried out in this chapter.

Many sequential runs (simulations) can be carried out consecutively if necessary. This option is useful if sensitivity analysis is to be carried out since using @RISK, values can be changed automatically from run to run.

The next significant part of this menu is the sampling type, since two options are offered in @RISK. The standard procedure for sampling from a probability distribution and the one used for all the simulations in this book so far, is the Monte Carlo method.

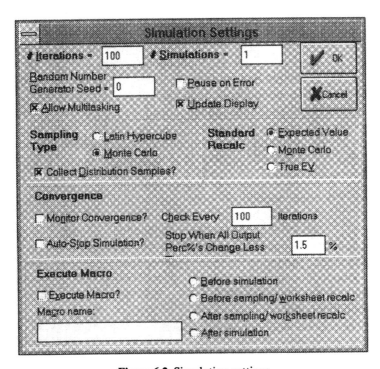

Figure 6.2 Simulation settings

Monte Carlo sampling is an entirely random process. Sampled values can be drawn from anywhere between the minimum and maximum values specified for a variable. Given a probability distribution, values are more likely to be sampled from those parts of the distribution that have the higher probabilities. A possible consequence of this, mentioned earlier, is that with smaller sample sizes low probability events in the tails of the input distributions will not be sampled. Examination of the sampled values of the input distributions, which it is possible to do, will reveal the extent of this and a judgement can be made about how serious a problem this might be.

Alternatively, a method known as Latin hypercube sampling may be used. This method uses a form of 'stratified' sampling. Each of the input distributions is divided into equal intervals or strata using its cumulative probability distribution. This is done according to the number of cycles or iterations, to be run. Thus if the number of samples to be taken is, as here, 500, then the CDF is divided into 500 equal (but small) strata.

A stratum is then selected randomly, and a value sampled from within that stratum. The sampling is done 'without replacement'. Once a value has been sampled from an interval, that interval is checked out of the distribution and is not sampled from again, because the stratum has been represented in the sample. Over the entire run therefore, every stratum is sampled once.

Among the advantages claimed for this method are:

1. A greater efficiency in sampling because fewer iterations are needed to be confident of representing all of an input distribution adequately.
2. Low probability events in the input variables are represented because the sampling process forces them to be included and their effects are present in the output distributions from the model.
3. Latin hypercube samples tend to converge to stable distributions more quickly than Monte Carlo samples, saving simulation time, especially with complex models.

Because the basic model is in effect a standard EXCEL spreadsheet, it may have to be used without an @RISK analysis. In this case, the cells that contain @RISK distribution formulae, such as D7 above, may need to contain a 'value' that can be used in any conventional calculation. The Standard Recalc button offers three options, although in most cases the default expected values option should be used. This puts the average, or expected value of the distribution into the cell, and the cell will thus contain a single value rather than the whole distribution.

The random number generator seed allows various levels of control over the first random number chosen. Use the same seed number for every run (e.g. 7121), and the same set of random numbers will be generated. This is most useful when the same model has to be run more than once, perhaps changing the shape of one of the probability distributions on each occasion. Then the same values will be sampled. Except for the changed distribution, any differences in the results from run to run can be attributed to the changes made. The default setting is zero. This causes a new random start number to be selected and a different series is generated for each run.

In Chapter 5 the question of stability was discussed. By monitoring 'convergence' the stability of the output can be tested. Changes in the basic statistics are tested by the system every 100 iterations. If the change is below a specified threshold the simulation is stopped because the statistics of the output distribution are sufficiently constant. The test should not of course be too coarse nor carried out on too small a sample since sufficient sample size remains a crucial factor.

6.3   @RISK OUTPUT

Generally, the output from @RISK is little different from that described in the Chapter 5 discussion. Basically this consists of summary statistics, and graphically a histogram and CDF. As was the case with the Chapter 5 examples, all of the combinations of variable values can be stored, displayed, printed and analysed. As before, as the sample size increases this can be very tedious and can usually be avoided without serious problems of interpretation.

An abbreviated set of statistics for a sample of 500 MSBs, using the three discrete distributions and the Monte Carlo sampling method, are shown in Table 6.3. In this table, the row 'description' indicates whether the variable is output, or an input probability distribution, which in this case is discrete. The row 'cell' shows the location of the variable used in

Table 6.3 Selected output from @RISK Monte Carlo sample: detailed statistics

Name	Maximum site bid (£): Monte Carlo sample	Change in building costs (%)	Cost of borrowing per calendar month (%)	Change in selling price per month (%)
Description	Output	Discrete(G4: G10,14:I10)	Discrete(K4: K14,M4;M14)	Discrete(O4: O14,Q4:Q14)
Cell	F46	D7	D9	D11
Minimum	387 411.1	3.00	0.50	0.50
Maximum	729 810.5	9.00	1.50	1.00
Mean	529 380.3	6.04	1.00	0.74
SD	60 532.49	1.481	0.251	0.125

the spreadsheet. Here the output variable, maximum site bid, is located in cell F46. The results can be compared almost directly with those in Table 6.1. Figures 6.3 and 6.4 can also be compared with Figure 6.1.

Statistics (and graphs) are also available to show the sample distributions of each of the three input variables, and these are also shown for this case in Table 6.3. These allow the user to observe how closely the

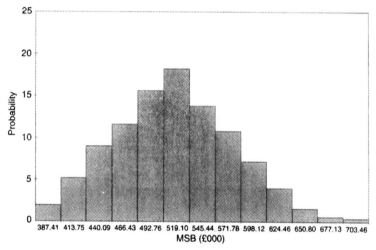

Figure 6.3 @RISK histogram for maximum site bid, 500 iterations: Monte Carlo sample

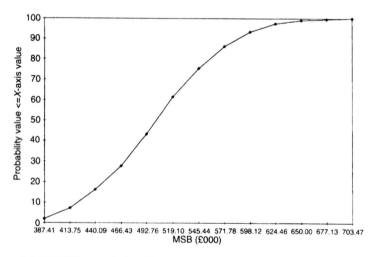

Figure 6.4 @RISK cumulative distribution for maximum site bid, 500 iterations: Monte Carlo sample

input distributions have been sampled. Table 6.3 shows for example, that the minimum value of the change in building cost variable that was sampled was 3%, the same as the lowest possible value. This indicates that sampling took place across the entire range of the variable.

By comparison with Table 6.1, the results from @RISK are very similar. Notice for example that the standard deviations are almost equal. Indeed, the coefficients of variation (0.114) are identical. For users of simulation models such as this, this is a comforting result in the sense that independent methods that are capable, by their nature, of producing different results, are not doing so to any significant extent.

Changing to the Latin hypercube sampling process for this appraisal, and running a simulation using this method, some subtle differences can be observed in Table 6.4.

As a consequence of the stratification the distribution is shifted and the standard deviation is increased because more observations have been sampled from the ends of the input distributions. In this sense the Latin hypercube sample is giving a rather more 'authentic' view of the overall risk in the appraisal since sampling has taken place across the entire range of each of the input distributions. This is also reflected in the 'flatter' shape of the histogram in Figure 6.5. Compared with Figure 6.3, observations are less concentrated in a single modal class, but the distribution drops away to the tails more quickly. (Although the graphs do not show this too well; in the case of the Monte Carlo sample 5% of the MSB distribution is above £630 000, with a maximum of £729 811). The Latin hypercube sample (Figure 6.6) has 5% of the distribution above £635 000, but with a lower maximum of about £707 380.

Table 6.4 Selected output from @RISK Latin hypercube sample: detailed statistics

Name	Maximum site bid (£): Latin Hyper-cube sample	Change in building costs (%)	Cost of borrowing per calendar month (%)	Change in selling price per month (%)
Description	Output	Discrete(G4: G10,14:I10)	Discrete(K4: K14,M4:M14)	Discrete(O4: O14,Q4:Q14)
Cell	F46	D7	D9	D11
Minimum	357 214.0	3.00	0.50	0.50
Maximum	707 380.2	9.00	1.50	1.00
Mean	531 975.1	6.00	1.00	0.75
SD	62 319.64	1.449	0.249	0.124

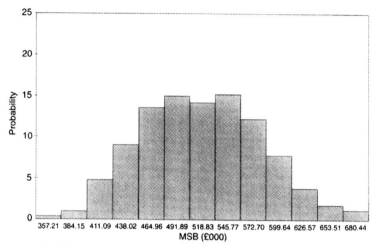

Figure 6.5 @RISK histogram for the maximum site bid, 500 iterations: Latin hypercube sample

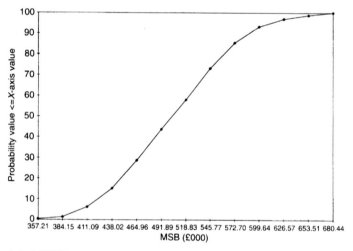

Figure 6.6 @RISK cumulative distribution for maximum site bid, 500 iterations: Latin hypercube sample

6.4 MODELLING THE INTERDEPENDENCE BETWEEN STATE VARIABLES

In Chapter 4, the problem of interdependence between state variables was briefly considered and one method for dealing with this was discussed. Interdependence is the extent to which one of the state variables is affected

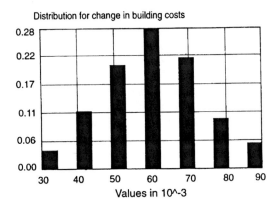

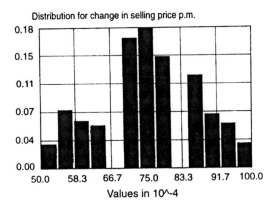

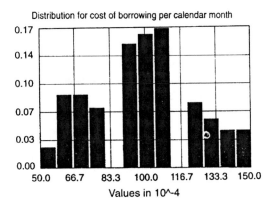

Figure 6.7 @RISK sampled distributions of correlated input variables, 500 iterations: Monte Carlo sampling

by another. For example, a high value in one would trigger a low value in another, and perhaps a high value in yet another. The method described can be applied, where the model is appropriate, by using @RISK to generate different shaped probability distributions to describe a state variable, dependent upon the value sampled for another state variable.

Here, because of the nature of the appraisal model that is being used, a different approach is adopted and demonstrated. This uses the level of correlation between the state variables. The assumption so far has been that the three state variables that are in our model are independent. Independence here means that a change in one would not cause, or be reflected by, a change in either or both of the others. In this case since they are independent, the correlation between the variables is zero, that is, there is no correlation. Correlation is a standard statistical concept. It simply measures the tendency of one variable to vary (linearly) with another. Correlation is measured as a coefficient on a scale from −1 through zero to +1.

Before considering the correlations that may exist among the three state variables, it is important to have clear a few more points about correlation.

A perfect positive correlation between two variables, A and B, gives a correlation coefficient of +1. As A increases so B increases in exact linear step with A. In contrast, a correlation coefficient of −1 implies a perfect negative correlation between two variables, A and B. That is, as A increases so B decreases in linear step with A, and vice versa. Intermediate values of the coefficient, suggest that the relationship is not as strong. For example, −0.5 indicates that when the value of A is high, the value of B will tend to be low, but not always. Thus, if two variables are strongly positively correlated then a high value in one should be matched by a high value in the other. This is what @RISK attempts to do during the sampling operation.

It is also important to remember two other important things about a correlation coefficient. First it is a purely numerical measure of a linear relationship. Other, more complex, non-linear kinds of relationship can exist between variables and assuming a linear relationship may give a misleading value. Secondly, the correlation between variables should be strongly based on theory, since numerically impressive correlation coefficients can be derived between variables that have no theoretical relationship at all. In preparing the correlations, what is to be avoided is a set of coefficients that are at odds in sign and size with what theory leads us to expect.

In this case therefore the first step has to be a consideration of the theoretical relationships between the state variables in our model. Generally it is held that when interest rates are high or increasing then housing starts tend to decline, mortgages are expensive and the demand for housing also tends to fall, depressing prices. Measuring these relationships consistently is hard to do, but this view implies negative correlations. As the borrowing rate rises, the other variables will tend to fall. If the decision-maker wanted a more exact view, this might well be obtained by examining for example, historic data relating the cost of borrowing to construction cost indices to see how the variables have performed relative to one another.

The correlation between borrowing rate and building cost is in fact more likely to be close to zero, and depending upon the exact state of the economy, no more than weakly negative or even positive. If interest rates are high, then costs may be kept low to encourage activity to be sustained. For this example, we assume a coefficient of –0.2.

If borrowing is high, then it is likely that change in house price will be low, and could be zero or could even have a negative rate of growth. This is not possible in our example when using the subjectively derived probability distributions. It is assumed that prices will rise by at least 0.5% per month. Even so, the correlation between the variables should be strongly negative, and here the value used is –0.8.

The correlation between building cost and house price is likely to be positive, and may be strongly positive. If building costs rise, this will be passed on to an extent in the asking price of the completed house. Of course there will be a trade off involved since this may result in houses remaining unsold, and extra borrowing will be incurred, reducing the overall profit to the developer. Here we assume a value of + 0.6.

It will be observed that the relationships can be complex. This is the principal reason why the issue of interdependence, although well appreciated, is often avoided in analyses using simulation methods. Once again, however, the ability to correlate state variables should encourage the user to examine the data that are available, and to question, or, more formally, theorize, as to the relationships between the variables. Then, in the absence of reasonable historic evidence on which to base the correlations, the user can estimate the coefficients, and if necessary test the model's sensitivity to the interdependence factor.

The correlation values are entered into a special correlation coefficients table or matrix in @RISK which looks like Table 6.5. Coefficients are mirrored above and below the diagonal in this table. On the diagonal each variable correlates perfectly, with itself.

Table 6.5 Correlation coefficients

	Change in building cost	Cost of borrowing	Change in selling price
Change in building cost	1	–0.2	0.6
Cost of borrowing	–0.2	1	–0.8
Change in selling price	0.6	–0.8	1

When the simulation is run @RISK uses a clever method to check that the results of sampling from each of the correlated variables are correctly related, in accordance with the correlation coefficients defined by the user.

The effects of this can be seen first in Figure 6.7. This shows the sampled distributions of the input variables that have resulted from the application of the correlation coefficients defined earlier. The shapes of the input distributions themselves are unchanged, but the shapes of the resulting sample distributions show clear differences, especially in borrowing and selling price because of the interdependency effects that are now being modelled. The building cost variable is treated as the baseline, and the other distributions are consequently bunched into the high and low values, as high and low values have been sampled from successive distributions. As Table 6.6 shows, however, the overall sample means and standard deviations of the input distributions are effectively little different to those sampled previously and shown in Tables 6.3 and 6.4.

There are, however, differences in the mean and standard deviation of the MSB. Both are higher than before, the SD significantly so. The correlation between the input variables seen in Figure 6.7, pulls the output values into the ends of the distribution. Although there is a strong modal element in the histogram (Figure 6.8), the rest of the distribution is quite flat, tailing away only gently. Clearly the ability to model this kind of interdependence in the simulation can have a marked effect on the kinds of results produced.

In the next section, the three input variables remain correlated using the values in Table 6.5. The correlation coefficients between the variables are unchanged and are independent of the types of probability distributions used. The correlations only affect the sampling process. In this section, however, continuous probability distributions are used for the input variables. Although they can be compared with each other, it is not possible to compare them directly with those that have been considered earlier, because these distributions have different statistical characteristics.

6.5 CONTINUOUS PROBABILITY DISTRIBUTIONS FOR STATE VARIABLES

It is often difficult, among many difficulties, to decide whether a state variable should be represented by a discrete or continuous probability distribution. In all of the analysis thus far, discrete forms have been used for the three state variables in the appraisal model. This is analytically very convenient, but it can be argued that often variables should really be modelled using continuous distributions. @RISK helps greatly by providing the complex mathematical and statistical specifications of a large

Table 6.6 Selected output from @RISK Monte Carlo correlated sample: detailed statistics

Name	Maximum site bid (£)	Change in building costs (%)	Cost of borrowing per calendar month (%)	Change in selling price per month (%)
Description	Output	Discrete(G4: G10,14:I10)	Discrete(K4: K14,M4:M14)	Discrete(O4: O14,Q4:Q14)
Minimum	373 055.0	3.00	0.50	0.50
Maximum	721 071.2	9.00	1.50	1.00
Mean	536 431.9	6.02	0.99	0.75
SD	74 754.58	1.436	0.250	0.122

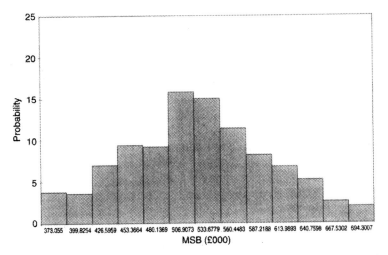

Figure 6.8 @RISK histogram for maximum site bid, 500 iterations: Monte Carlo sample, discrete input distributions, correlated

number of continuous probability distributions. These would otherwise be very tedious to model.

The use of continuous distributions to model state variables brings other changes in methodology as well. Some of the analysis is beyond the scope of this book. Here three continuous probability distributions are utilized to show the main features of simulation using such distributions.

The most significant feature of a variable that is modelled by a continuous distribution is that it may take any value within the range defined by the parameters and the mathematical function that describe a particular distribution. Some of the general points relating to the use of continuous distributions can be discussed by reference to the most commonly employed of these distributions, the normal distribution.

The normal distribution

The characteristics of this continuous probability distribution, including the function used to calculate its shape, are described in some detail in the appendix. It has been observed that many naturally arising random variables can be reasonably approximated by this distribution. In consequence there has been a quite sensible tendency to use the distribution to describe other more artificially derived variables, and it is a very important statistical distribution. The main features of the distribution are that it is symmetrical and open ended. This symmetry means that the mean (expected) value, the median and mode are all at the same, central, position in the distribution. If it is possible to estimate a 'most likely' value for the variable, the mean would take the same value. The spread of the distribution is described by the standard deviation (SD), and although it is open-ended almost all of the distribution lies within ± 3 standard deviations of the mean. Most importantly, any normal distribution can be described completely using only its mean and SD. This means that, as long as the assumption that a variable is normally distributed is acceptable, there is no need to construct a subjective probability distribution, since estimates of the mean and SD are all that is required to be able to define a normal distribution for an @RISK simulation. Clearly there is still a need to decide how the uncertainty in each of the state variables is to be assessed and described. Earlier it was suggested that the construction of complete subjective distributions can be a valuable part of the decision-making process by drawing the appropriate experts more deeply into the appraisal process. On the other hand, if a decision is made to assume a normal distribution, then the problem is a bit different. Essentially, the mean/SD relationship has to reflect the uncertainty that is associated with that variable.

To show the effect of this, the variable 'change in building cost' has been used to generate two very large samples based on the assumption that the variable is normally distributed. In the first case, the mean is 0.06, and the SD, 0.01. In the second case, the SD is doubled in size to 0.02, and the uncertainty associated with that variable increases accordingly. This is in fact not an unreasonable assumption in that a 'rate of change' variable might be expected to take any value along a continuous scale, subject to an ability to measure it and any limits to the variable.

Here, for each of the normal distributions, input to @RISK consists of the commands:

RiskNormal (0.06,0.01)
RiskNormal (0.06,0.02)

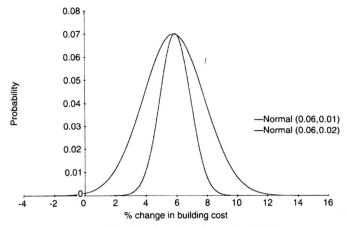

Figure 6.9 Different normal distributions for change in building cost generated by @RISK

The first value in the bracket is the estimated mean, and the second the SD.

The results can be seen in Figure 6.9. Statistically, both distributions are normal. Even so, although the sample size used to generate these distributions is very large, the effects of sampling can still be observed. In theory these two distributions should show the same mean (0.06), but there is obviously a small difference in the mean of each of the distributions.

Each of the state variables is now to be represented by a normal distribution, defined by a mean and an SD as follows:

1. RiskNormal (0.06,0.01)   represents change in building cost (this is the distribution shown in Figure 6.9).
2. RiskNormal (0.009,0.001)   represents cost of borrowing.
3. RiskNormal (0.0075,0.001) represents change in selling price.

Note that the values used for the parameters of the continuous distributions are not necessarily the same as those used in Chapter 5.

The simulation is then run, using the Monte Carlo method in this case. The results and the statistics of the input variables are shown in Table 6.7.

The distributions of the input variables are open-ended and are not now restricted to the discrete end-points which were previously defined. Because of the values of the SDs that are being used, however, the maxima and minima that emerge are quite similar to those observed earlier.

The resulting histogram, shown in Figure 6.10, shows that the output distribution is not by any means symmetrical. It also shows a fairly strongly modal value. In this way both the normal distributions and the correlations are reflected to some extent in the output. The SD is reduced,

Table 6.7 Correlated input variables with normal distributions: @RISK detailed statistics

Name	Maximum site bid (£)	Change in building costs	Cost of borrowing per calendar month	Change in selling price per month
Description	Output	Normal(G7,0.01)	Normal(K8,0.00)	Normal(O9,0.001)
Minimum	395 478.0	0.0332	0.0063	0.0036
Maximum	681 445.6	0.0885	0.0125	0.0104
Mean	550 211.1	0.0600	0.0090	0.0075
SD	44 450.62	0.0098	0.0010	0.0010

Note: The input values are expressed here in decimal rather than percentage form.

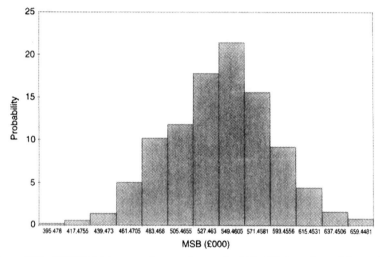

Figure 6.10 @RISK histogram for maximum site bid, 500 iterations: Monte Carlo sample, normal input distributions, correlated

matching the relatively narrow (certain) normal distributions used to model the input variables.

When a change is made to the SDs, there is a substantial increase in the output SD, as would be expected. The SDs of the cost of borrowing and the change in selling price variables are now each doubled in size. The SD of the change in building cost is unchanged. The means of all three variables stay the same. The input specifications for the variables are now:

1. RiskNormal (0.06,0.01)
2. RiskNormal (0.009,0.002)
3. RiskNormal (0.0075,0.02)

Table 6.8 Correlated input variables with normal distributions: @RISK detailed statistics

(Increased input standard deviations)

Name	Maximum site bid (£)	Change in building costs	Cost of borrowing per calendar month	Change in selling price per month
Description	Output	Normal(G7,0.01)	Normal(K8,0.002)	Normal(O9,0.002)
Minimum	251 649.0	0.0332	0.0037	−0.0003
Maximum	842 140.1	0.0885	0.0159	0.0132
Mean	554 431.1	0.0600	0.0089	0.0075
SD	92 899.31	0.0098	0.0020	0.0020

These values can be seen in Table 6.8. Because the same random number sequence is being used, the statistics of the change in building cost variable do not show any variation at all. All of the observed differences can be attributed only to the two changed input SDs. A particular point to notice here is that a variable which could only take a positive value when using other input distributions, change in selling price, now has a minimum which is just negative, equivalent to −0.30% per annum.

As a consequence of these alterations, the minimum MSB falls from £339 548 to £251 649, and the maximum rises from £682 446 to £842 140 (Figure 6.11). The mean of the sample hardly changes at all, but the SD is more than doubled rising from 44 451 to 92 899.

Although the normal distribution has many advantages, it also has

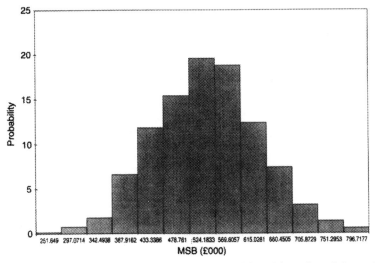

Figure 6.11 @RISK histogram for maximum site bid, 500 iterations: Monte Carlo sample, normal input distributions, correlated, high variance

some limitations. The most significant of these are unfortunately that it is symmetrical and open-ended. The discrete variables used for the simulations in Chapter 5 and earlier in this chapter are symmetrical, but are not open-ended. When the subjective distributions were constructed in Chapter 4, they were clearly not open-ended, nor were they particularly symmetrical. Part of the process of constructing the distributions was to define values as the end-points beyond which there was no probability of occurrence. This does not happen with the normal distribution. There is always a very small probability that a value will be observed from the, theoretically infinite, tail(s) of a normal distribution. Equally, the shapes of the probability distributions of economic variables are not symmetrical, but are very often skewed to a greater or lesser extent. Because of this, interest has been directed at some other distributions that can deal with these problems.

The triangular distribution

A 'continuous' distribution that is sometimes suggested as a partial improvement on the normal, but which is easy to describe, is the triangular. The triangular distribution is closed, and can take symmetrical or skewed forms. The distribution needs three measures to describe its form completely. These are a minimum value, a most likely value and a maximum value. The shape of a typical triangular distribution is shown in Figure 6.12. In this case $a = 0$, $b = 1$ and $c = 10$.

Because of its form, the triangular distribution is only approximating the continuous relationship between the variable and its associated probability, since the minimum, most likely and maximum values are connected by straight lines. It can be used, however, to approximate a skewed distribution, and it has the distinct advantage of still only requiring the three estimates to enable it to be constructed. It is also particularly useful in that the values required to estimate the distribution are the absolute minimum

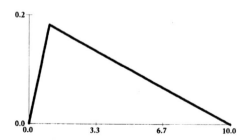

Figure 6.12 A typical triangular distribution

that any adviser might be reasonably expected to be capable of providing. If they could not then their 'expertise' might be strongly questioned!

The mean of a triangular distribution is approximated by:

$$m_T = \frac{a + b + c}{3}$$

where: a = minimum, b = most likely and c = maximum value.

and the variance is approximately:

$$v_T = \frac{a^2 + b^2 + c^2 - ab - ac - bc}{18}$$

As with the normal distribution, although the effort of constructing a complete probability distribution need not be made in any detail, there is still the need to decide positively what the parameters of the distribution should be. The values of a, b and c have been chosen deliberately to produce moderately skewed distributions. The resulting triangular input distributions are shown in Figure 6.13. The values chosen are entered into the @RISK function for the triangular distribution: RiskTriang(minimum, mode, maximum):

1. RiskTriang (3, 5, 9)
2. RiskTriang (0.5, 0.9, 1,5)
3. RiskTriang (0.5, 0.65, 1.0)

When the simulation is run, Table 6.9 shows that even though the distributions of the input variables are bounded by maximum and minimum values, the Monte Carlo method in this example does not sample at the exact ends of the distributions, except in the one case of the lower boundary of the change in selling price variable.

The output results seen here are not out of line with those seen from other distributions. The SD is rather high, but is less than that seen in the correlated discrete case and more than in the normal case (Tables 6.6 and 6.7) before the SDs were substantially increased (Table 6.8).

Figure 6.14 shows that the triangular input distributions produce what

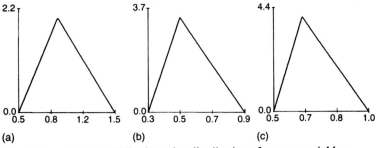

Figure 6.13 @RISK triangular distributions for state variables

Table 6.9 Correlated input variables with triangular distributions: @RISK detailed statistics

Name	Maximum site bid (£)	Change in building costs	Cost of borrowing per calendar month	Change in selling price per month
Description	Output	Triang (G4,G6,G10)	Triang (K4,K8,K14)	Triang (O4,O7,O14)
Minimum	390 336.2	0.0321	0.0054	0.0050
Maximum	686 616.0	0.0877	0.0149	0.0098
Mean	533 021.1	0.0566	0.0096	0.0072
SD	61 713.35	0.0124	0.0020	0.0010

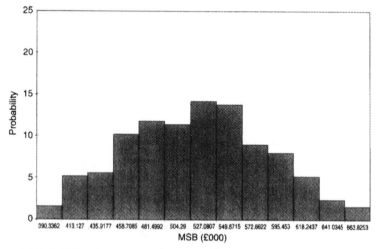

Figure 6.14 @RISK histogram for maximum site bid, 500 iterations: Monte Carlo sample, triangular input distributions, correlated

is perhaps the flattest of the MSB distributions. Although it does have recognizable modal classes, the rest of the distribution is relatively evenly spread across the histogram, declining only gradually into the tails. This is almost certainly due to the combination of three interacting effects, the correlation between the input variables, the skewed nature of the input distributions, and the over-estimation of the variance which is known to be a problem with the triangular distribution.

The triangular distribution is a quick and dirty approximation. If a system such as @RISK is available, then a better choice, using the same basic measures, is the beta distribution. The beta will amongst other things overcome the tendency of the triangular distribution to overestimate the variance. Deriving an appropriate beta distribution is, however, somewhat more complicated than in the triangular case.

The beta distribution

The beta is a truly continuous distribution. Although it is has finite endpoints, a minimum and a maximum, it can take a very wide variety of shapes, because the modal value can be placed anywhere within the distribution.

For many years the most common use of the beta distribution has been in the program evaluation and review technique (PERT). PERT is a method of project scheduling. It is similar to critical path analysis (CPA), which is often used in the construction industry for project planning. The crucial difference is that PERT explicitly includes risk measures of the likely time that each of the activities in a project might take. These times for completion are usually modelled on the assumption that they take a beta distribution. The beta can of course be used to model other random variables.

The Beta distribution requires two parameters, called $\alpha 1$ and $\alpha 2$ (alpha1 and alpha2) in @RISK, to define the shape of the distribution. To show the variety of shapes which are possible with this distribution, dependent upon the values of $\alpha 1$ and $\alpha 2$, Figures 6.15(a) and (b) show beta distributions with $\alpha 1 = 1$ and $\alpha 2 = 3$, and with $\alpha 1 = 3$ and $\alpha 2 = 2$.

How are the $\alpha 1$ and $\alpha 2$ parameters found, so that the appropriate beta is used in the simulation? The first step is to obtain the same three estimates for the distribution as in the triangular case, a minimum value (a), a most likely value (b) and a maximum value (c). Here they have the same values as were used in the triangular case.

Given these values then a mean (m) and variance (v) can be estimated for the beta:

$$m = \frac{a + 4b + c}{6}$$

and

$$v = \left(\frac{c - a}{6}\right)^2$$

In calculating the mean, the mode (or most likely value) is weighted four times more heavily than the other values. For the variance calculation, the rationale, as with the normal distribution, is that the probability density function concentrates almost all of the distribution within the three standard deviations on either side of the mean. There are thus six standard deviations between a and c.

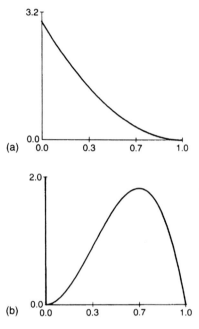

(a)

(b)

Figure 6.15 (a) Beta distribution with alpha1 = 1 and alpha2 = 3; (b) beta distribution with alpha1 = 3 and alpha2 = 2

Using these estimates, the values of α1 and α2 can be derived using the equations below:

$$\alpha 1 = \left(\frac{m-a}{c-a}\right)^2 * \left(1 - \frac{m-a}{c-a}\right) * \left(\frac{v}{(c-a)^2}\right)^{-1} - \frac{(m-a)}{(c-a)}$$

$$\alpha 2 = \left(\frac{m-a}{c-a}\right) * \left(1 - \frac{m-a}{c-a}\right) * \left(\frac{v}{(c-a)^2}\right)^{-1} - \left(1 - \frac{(m-a)}{(c-a)}\right)$$

These may seem to be complicated equations, but the spreadsheet will solve them easily. Using the same values for *a*, *b* and *c* as were used in the triangular distributions earlier, and these formulae, the α1 and α2 values for the three state variables can be found and they are summarized in Table 6.10.

The parameters of the beta distributions for each of the variables are now defined. The resultant shape of the beta distribution for change in building cost, with α1 = 2.938 and α2 = 4.617 is shown in Figure 6.16.

@RISK samples from a standard form of the beta calculated in terms of the values α1 and α2. After sampling from this standard form of the beta, the value obtained has to be transformed to the 'real' scale of each variable. This is done by inserting the following formula in the appropriate cell(s):

Table 6.10 Parameters and alpha values for Beta input variables

	Change in buildings costs	Cost of borrowing per calendar month	Change in selling price per month
Minimum	3.000	0.500	0.500
Mode	5.000	0.900	0.650
Maximum	9.000	1.500	1.000
Mean	5.333	0.933	0.683
Variance	1.000	0.028	0.007
alpha 1	2.938	3.397	2.699
alpha 2	4.617	4.443	4.661

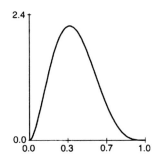

Figure 6.16 Beta distribution for change in building cost

Minimum + (RiskBeta(alpha1,alpha2) * (Maximum-Minimum))

The 'results' for each of the variables are then used as sampled values in one cycle of the simulation.

Table 6.11 shows the effect of the reduced variance in the input beta distributions, by comparison with their triangular equivalents. This is seen in the SD of the MSB, which is reduced to 49 514, compared with 61 713 in Table 6.8. Obviously this also results in the reduced spread seen in the histogram (Figure 6.17). This histogram is, however, much more regular in shape than Figure 6.14 and this is the result of smooth beta distributions used for the input variables, in contrast with the angular 'approximation' of Figure 6.13.

The choice as to which probability distribution to use for any variable is not necessarily easy. Very often 'it is assumed that the distribution is normal' to make life analytically simple. The development of a discrete distribution by subjective probability methods is the best course of action if the time and effort involved can be expended, and even if the result is

Table 6.11 Correlated input variables with Beta distributions: @RISK detailed statistics

Name	Maximum site bid (£)	Change in building costs	Cost of borrowing per calendar month	Change in selling price per month
Description	Output	Beta (2.938,4.617)	Beta (3.397,4.443)	Beta (2.699,4.661)
Minimum	402 743.2	3.302	0.573	0.504
Maximum	661 311.4	8.096	1.422	0.920
Mean	529 980.6	5.327	0.926	0.685
SD	49 514.0	0.989	0.166	0.081

Note: The input variable values are expressed here in percentage form.

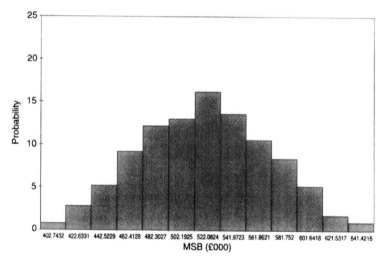

Figure 6.17 @RISK histogram for maximum site bid, 500 iterations: Monte Carlo sample, beta input distributions, correlated

approximate. The overall information gathering power of the procedure is large, and this often outweighs the loss of statistical rigour which continuous distributions impose because their functional form can be employed to calculate accurate probabilities for given variables. With continuous variables, in the absence of sufficient data to be able to estimate what the most likely form of distribution might be, then rules of thumb are best applied. Bearing in mind that there are many continuous distributions, then, use the normal as an approximation if the variable is felt to be reasonably symmetrically distributed, even though it is open-ended. Use the beta if the variable is thought to be skewed to either the left or right.

6.6 SIMULATION IN THE DEVELOPMENT APPRAISAL PROCESS

Chapters 5 and 6 have taken a relatively simple appraisal problem and subjected it to a probabilistic, cash flow simulation. The Monte Carlo and Latin hypercube methods have been applied to sample the values of a number of logically probabilistic variables. Use has been made of both discrete and continuous distributions to model those variables and interdependence has been built into the model.

Probabilistic simulation is essentially a computer based technique. Simulations of the kind demonstrated in this book clearly require a large number of calculations to be made, both in the sampling of variable values by either of the Monte Carlo or Latin hypercube methods, then in applying those values in the appraisal model, and finally in repeating the process a relatively large number of times. Modelling of any kind can be a powerful tool, but the ability to sample from probability distributions is a very important element, making it possible to allow for uncertainties in complex, risk-bearing, problems.

Applied to appraisal problems, the use of this methodology has reemphasized a number of requirements that have been separately discussed, but which are brought together in the simulation approach:

1. Using the cash flow approach as the basis for appraisal, the pattern of cash flows has to be defined explicitly.
2. The values of each of the variables are tested logically as either deterministic or probabilistic.
3. The performance of the probabilistic variables has to be assessed.
4. The resulting cumulative distributions have to be tested for stability and sensitivity.

These analyses, which must precede the simulation, can yield a very large amount of information about the problem, and properly carried out provides consistent input data for the simulation.

As with any technique of this kind, the final results will only be as good as the initial data that go into the model. Time, energy and money are necessary expenditures to achieve as high a quality of 'hard' data as possible. This is true, of course, not only for the simulation, but for each analysis made in the decision process. The basic requirements are that there be

1. A proper understanding of the assumptions and limitations to be applied to the data, and

2. that the methodology itself is not extended beyond its capabilities, thus invalidating the results of what might otherwise be an acceptable, and valuable analysis.

NOTES

* @RISK is a Risk Analysis and Simulation Add-In for Microsoft EXCEL and LOTUS 123. The version used here is Release 3.0 for Windows. It is produced and copyrighted by Palisade Corporation, 31 Decker Road, Newfield, NY, USA, 14867.

Postscript

Only a very small proportion of individuals positively like the challenge of having to deal with risk and uncertainty on a day-to-day basis. It is for this reason that, in the last twenty to thirty years, considerable research effort has gone into the development of methods and approaches that are designed to help in dealing with this challenge, and to minimize the disturbance that these factors so often cause.

In general these approaches have gained only a slow acceptance, although some techniques have been taken up more readily, to be used in problems where they have been felt to be particularly appropriate, as in this book.

A number of factors can be identified as having inhibited the wider use of formal methods of decision analysis (Moore, 1977).

The direct measurement of probability (and utility) is, by its nature, alien to most decision-makers. As has been shown, this is remarkable in the sense that many seem quite able to apply the underlying concepts quite happily when making 'ordinary' decisions, but are prepared to reject, almost out of hand, any procedure which imposes a structure on an otherwise commonly occurring, but intuitive, process. Related to this is the difficulty which decision-makers appear to have in believing that it is possible for probabilities to be assessed differently by various individuals depending upon their perceptions of possible outcomes to a particular problem. Even if probabilities are measured in ways which can be appreciated by the decision-maker, their application to derive expected values (on monetary or utility scales) can, in turn, present problems. This is simply because the expected value only ever represents an average measure of problem performance, good for comparative purposes; but it only rarely has a 'real' value capable of being interpreted in absolute terms. A further difficulty cited by decision-makers is that the 'intangibles' which exist in all decisions nullify systematic decision methods because they cannot be measured, and cannot therefore be included in the analysis. Strictly, this can indeed be the case, but the major benefit comes from the attention focused on these intangible elements. From this it may well be possible to describe the extent to which they are 'bounded', even if only on a subjective basis.

Some of these difficulties can be removed by the continuing process of education and training which will lead to a much greater understanding and appreciation of these methods, and hence to the circumstances in which they are of value to particular decision-makers, including those concerned with property development and investment.

Even with the wider acceptance of these basic concepts, coherent and consistent decisions can really only be made when the maximum amount of relevant information is available. Information remains the fundamental factor. Without it, and because of resulting uncertainty, wrong avenues are easily followed, or good propositions are rejected, no matter what methods of analysis are employed. The more coherent the basic methods employed, the more consistent will decisions be, even in property analysis.

In recent years much work has been carried out with a view to making this possible. This effort has gone a fair way towards removing the considerable variety in the way that property is analysed. There are continuing attempts to standardize the definitions of the various processes of valuation, in particular to systematize the methods of conventional property investment valuation, and to understand how risk and uncertainty impinge on property as an investment. These developments may not affect the process of property development directly, but since many of those who advise developers are valuers as well, the advances in valuation methodology may well be expected to percolate ultimately into all aspects of the property analyst's work.

There has also been considerable research and development of property performance measurement systems. These are designed primarily to analyse portfolios of existing property, but equally they can provide, for the first time, the possibility of consistent measurement bases for comparing new development/investment returns against those being curently obtained.

Even within the constraints of commercial confidentiality, much more, and harder data are becoming available in those areas of property analysis where good information matters most.

In the specific areas of risk and uncertainty the substantial changes have really been in the technology, in the hardware and software that make the analysis possible. The processing power of current business computers is easily sufficient to handle any of the kinds of analysis discussed in this book in almost real time, and it is available at a reasonable cost for both hardware and software. This computing power is especially valuable in terms of the treatment of sensitivity, since the decision-maker can see instantly the likely effects of changes in crucial variables in any decision model. In terms of methodological innovations the most recent and certainly the most radical of these is the possibility of using the methods of fuzzy logic as an alternative to probability-based analysis. This is still even more experimental than the use of simulation type methods, but offers another potential option for the decision-maker (Byrne, 1995).

Taken together with the methods for dealing more directly with risk and uncertainty, these developments continue to hold out the promise of better, more accurate, and more consistently profitable decisions in property even under the worse conditions of risk and uncertainty!

Appendix: The characteristics of the normal distribution

Many randomly occurring processes or variables have their events measured on a continuous scale, examples being the Financial Times Share Index, or the value of the pound against the dollar. Each of these is measured on a quite fine scale and can take large numbers of values between rather wide limits.

By comparison, most of the calculations made in this book assume that the values of the random variables are separately classified, i.e. $x = 1, 2$ or 3, with no intermediate measurement. These variables are called discrete.

In the continuous case, the likelihood of occurrence of various outcomes is calculated using a probability density function rather than a discrete probability distribution. The probability density function implies that there is some equation which will enable the probability to be calculated for any value of the random variable, and from this equation the overall shape of the continuous probability distribution can be derived. All that needs to be known for this to be possible are the parameters which go into the equation for any particular density function. Many random variables may have the same basic density function, but may differ in the value of their parameters. They then belong to one family of distributions. One such family of random variables is described by the normal distribution.

The normal distribution is one of the most important of the continuous probability distributions. Practically, the normal distribution closely fits the observed frequency distributions of many phenomena, and many others can be assumed to take this distribution, even if the actual distribution has not been observed.

The general shape of the curve has been seen in figures in Chapter 6

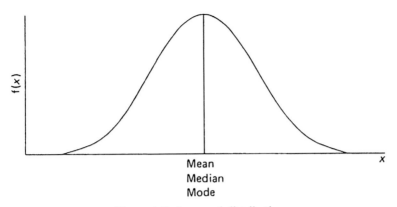

Mean
Median
Mode

Figure A.1 A normal distribution

but is shown in Figure A.1. The curve is shaped like a bell, with a single symmetrical peak. The mean, median and mode lie at the centre of the normal curve. The function for the normal curve is such that the curve is open-ended at both ends, gradually approaching the horizontal axis but never reaching it.

The equation for the normal curve is:

$$f(x) = \frac{1}{\sqrt{2\pi\sigma^2}}\, e^{-\frac{(x-\mu)^2}{2\sigma^2}}$$

where $\pi = 3.1416$ and $e = 2.71828$, μ is the mean and σ the standard deviation of the probability distribution, and x is any value of the random variable X.

This is obviously complicated to calculate and it is usually obtained by reference to a prepared table of values, such as that given in Table A.1.

Within the general family shape, the actual shape of any normal distribution will be governed by the standard deviation. Distributions with small standard deviations are very 'peaky', and those with large standard deviations are flatter, although they will still be distinctly uni-modal.

PROBABILITY AND THE NORMAL DISTRIBUTION

For any random variable the height of the curve is the probability density function $f(x)$, where x is a possible value of the random variable X. The total area under the (theoretically infinite) curve is 1.00.

The probability that a normally distributed random variable has values between any interval of X is equal to the area of the portion of the curve

Table A.1 Areas under the normal curve

Z	0	0.01	0.02	0.03	0.04	0.05	0.06	0.07	0.08	0.09
0	0.0000	0.0040	0.0080	0.0120	0.0160	0.0199	0.0239	0.0279	0.0319	0.0359
0.10	0.0398	0.0438	0.0478	0.0517	0.0557	0.0596	0.0636	0.0675	0.0714	0.0753
0.20	0.0793	0.0832	0.0871	0.0910	0.0948	0.0987	0.1026	0.1064	0.1103	0.1141
0.30	0.1179	0.1217	0.1255	0.1293	0.1331	0.1368	0.1406	0.1443	0.1480	0.1517
0.40	0.1554	0.1591	0.1628	0.1664	0.1700	0.1736	0.1772	0.1808	0.1844	0.1879
0.50	0.1915	0.1950	0.1985	0.2019	0.2054	0.2088	0.2123	0.2157	0.2190	0.2224
0.60	0.2257	0.2291	0.2324	0.2357	0.2389	0.2422	0.2454	0.2486	0.2517	0.2549
0.70	0.2580	0.2611	0.2642	0.2673	0.2704	0.2734	0.2764	0.2794	0.2823	0.2852
0.80	0.2881	0.2910	0.2939	0.2967	0.2995	0.3023	0.3051	0.3078	0.3106	0.3133
0.90	0.3159	0.3186	0.3212	0.3238	0.3264	0.3289	0.3315	0.3340	0.3365	0.3389
1.00	0.3413	0.3438	0.3461	0.3485	0.3508	0.3531	0.3554	0.3577	0.3599	0.3621
1.10	0.3643	0.3665	0.3686	0.3708	0.3729	0.3749	0.3770	0.3790	0.3810	0.3830
1.20	0.3849	0.3869	0.3888	0.3907	0.3925	0.3944	0.3962	0.3980	0.3997	0.4015
1.30	0.4032	0.4049	0.4066	0.4082	0.4099	0.4115	0.4131	0.4147	0.4162	0.4177
1.40	0.4192	0.4207	0.4222	0.4236	0.4251	0.4265	0.4279	0.4292	0.4306	0.4319
1.50	0.4332	0.4345	0.4357	0.4370	0.4382	0.4394	0.4406	0.4418	0.4429	0.4441
1.60	0.4452	0.4463	0.4474	0.4484	0.4495	0.4505	0.4515	0.4525	0.4535	0.4545
1.70	0.4554	0.4564	0.4573	0.4582	0.4591	0.4599	0.4608	0.4616	0.4625	0.4633
1.80	0.4641	0.4649	0.4656	0.4664	0.4671	0.4678	0.4686	0.4693	0.4699	0.4706
1.90	0.4713	0.4719	0.4726	0.4732	0.4738	0.4744	0.4750	0.4756	0.4761	0.4767
2.00	0.4772	0.4778	0.4783	0.4788	0.4793	0.4798	0.4803	0.4808	0.4812	0.4817
2.10	0.4821	0.4826	0.4830	0.4834	0.4838	0.4842	0.4846	0.4850	0.4854	0.4857
2.20	0.4861	0.4864	0.4868	0.4871	0.4875	0.4878	0.4881	0.4884	0.4887	0.4890
2.30	0.4893	0.4896	0.4898	0.4901	0.4904	0.4906	0.4909	0.4911	0.4913	0.4916
2.40	0.4918	0.4920	0.4922	0.4925	0.4927	0.4929	0.4931	0.4932	0.4934	0.4936
2.50	0.4938	0.4940	0.4941	0.4943	0.4945	0.4946	0.4948	0.4949	0.4951	0.4952
2.60	0.4953	0.4955	0.4956	0.4957	0.4959	0.4960	0.4961	0.4962	0.4963	0.4964
2.70	0.4965	0.4966	0.4967	0.4968	0.4969	0.4970	0.4971	0.4972	0.4973	0.4974
2.80	0.4974	0.4975	0.4976	0.4977	0.4977	0.4978	0.4979	0.4979	0.4980	0.4981
2.90	0.4981	0.4982	0.4982	0.4983	0.4984	0.4984	0.4985	0.4985	0.4986	0.4986
3.00	0.4987	0.4987	0.4987	0.4988	0.4988	0.4989	0.4989	0.4989	0.4990	0.4990

covering the interval. This can be calculated using the function for the normal curve, but this is tedious and is usually tabulated for a particular normal distribution, the standard normal distribution.

The standard normal distribution is identified as having a mean of zero and a standard deviation of unity, the random variable of this special case being identified as Z rather than X.

It can then be shown from the equation of the normal curve that:

$$P(0 \leqslant Z \leqslant 1) = 0.3413$$

$$P(0 \leqslant Z \leqslant 2) = 0.4772$$

$$P(0 \leqslant Z \leqslant 3) = 0.4987$$

Since it is a symmetrical distribution:

$$P(-1 \leqslant Z \leqslant +1) = 0.6826$$

$$P(-2 \leqslant Z \leqslant +2) = 0.9544$$

$$P(-3 \leqslant Z \leqslant +3) = 0.9974$$

The standard deviation of this distribution is unity, and so it can be said that 68.26% of the values of the random variable Z will fall within ± 1 standard deviation of the mean, 95.44% fall between ± 2 standard deviations from the mean, and 99.74% fall between ± 3 standard deviations from the mean. This view may be extended indefinitely into either tail of the distribution, but this is rarely done since the proportion of observations in the tails is very small.

These proportions can be shown to hold for any normal distribution, for any values of the mean and standard deviation, and are summarized in Figure A.2.

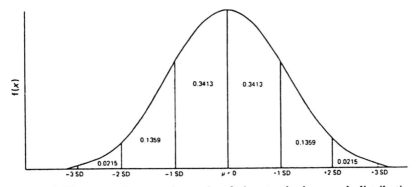

Figure A.2 The areas between intervals of the standard normal distribution: $\mu = 0$, $\sigma = 1$

Using the tables of the normal distribution *any* area (probability) can be found. For example:

1. What is the probability that a standard normal random variable takes values between $+2$ and $+3$?

    $$P(+2 \leqslant Z \leqslant +3) = P(Z \leqslant +3) - P(Z \leqslant +2)$$

    From above:

    $$P(Z \leqslant +3) = 0.4987$$
    $$P(Z \leqslant +2) = 0.4772$$
    $$P(+2 \leqslant Z \leqslant +3) = 0.4987 - 0.4472 = 0.0215$$

2. What is the probability that the variable takes values between -1 and $+1.5$?

    $$P(-1 \leqslant Z \leqslant +1.5) = P(Z \leqslant +1.5) - P(Z \leqslant -1)$$

    From tables:

    $$P(Z \leqslant +1.5) = 0.4332$$
    $$P(Z \leqslant -1) = -0.1359$$
    $$P(-1 \leqslant Z \leqslant +1.5) = 0.4332 - (-0.1359) = 0.5691$$

Tables prepared for the standard normal distribution can be used for any normally distributed random variable X. This is essential since most random variables of this kind do not have means of zero and standard deviations of unity.

To demonstrate this consider an investment which produces an average return of 9% with a standard deviation of 2%. The distribution of returns is normal. How can the probability that the yield will be between 10% and 13% be found?

In general any normally distributed random variable X can be converted to the standard normal Z by:

$$Z = \frac{x - \mu}{\sigma}$$

where μ and σ are the mean and standard deviation respectively of the random variable X, and x is any value of that variable. Here $\mu = 9$, $\sigma = 2$ and $x = 10$ and 13. Then:

$$Z = \frac{10 - 9}{2} = 0.5$$

$$Z = \frac{13 - 9}{2} = 2.0$$

So that:

$P(10 \leqslant X \leqslant 13) = P(0.5 \leqslant Z \leqslant 2.0)$

From the tables:

$P(Z \leqslant 2.0) = 0.4772$

$P(Z \leqslant 0.5) = 0.1915$

Then:

$P(10 \leqslant X \leqslant 13) = P(X \leqslant 13) - P(X \leqslant 10) = 0.2857$

Bibliography

This bibliography, as well as giving the references used in the text, is intended to show the considerable depth and breadth of work which has been done in the field of decision-making under conditions of risk and uncertainty both generally and, to a more limited extent, in property. It is not exhaustive! These references extend the treatment of subjective probability, utility and simulation methods introduced in this book.

The references are for the most part at the same, or only marginally more, mathematical level as is employed in this book. A few are, however, very technical but are included to provide a more complete coverage of the topic. Still more depth can be found by looking at further references derived in their turn from those given here.

Arnison, C. and Barrett, A. (1985), Valuations of development sites using the stochastic decision tree method. *Journal of Property Valuation*, 3 (2), 126–133

Barras, R. (1979), *The Development Cycle in the City of London*, Research Series 36. CES Ltd, London.

Barras, R. (1994), Property and the economic cycle: building cycles revisited. *Journal of Property Research*, 11 (3), 183–197.

Baum, A. E. (1978), Residual valuations: a cash flow approach. *Estates Gazette*, 247, 973–975.

Baum, A. E. and Mackmin, D. H. (1981), *The Income Approach to Property Valuation*, 2nd edn, Routledge and Kegan Paul, London.

Bierman, H. and Smidt, S. (1988), *The Capital Budgeting Decision*, 7th edn, Collier Macmillan, West Drayton.

Bonini, C. P. (1975), Risk evaluation of investment projects. *Omega*, 3 (5), 737–750.

Byrne, P. (1995), Fuzzy analysis: a vague way of dealing with uncertainty in real estate analysis? *Journal of Property Valuation and Investment*, 3 (13), 3, 22–40.

Cadman, D. (1984), Property finance in the UK in the post-war period. *Land Development Studies*, 1 (2).

Cadman, D. and Catalano, A. (1983), *Property Development in the UK: Evolution and Change*, Centre for Advanced Land Use Studies, Reading (available from E. & F. N. Spon, London).

Carter, E. E. (1972), What are the risks in risk analysis? *Harvard Business Review*, 50, 72–82.

Corgel, J. B. (1980), On improving interpretation of simulated investment values. *Real Estate Appraiser and Analyst*, 46 (6), 16–22.

Coyle, R. G. (1972), *Decision Analysis*, Nelson, London.

Curcio, R. S., Gaines, J. P. and Webb, J. R. (1981), Alternatives for assessing risk in real estate investments. *Real Estate Issues*, 6 (2), 25–32.

Darlow, C (ed.) (1988), *Valuation and Development Appraisal*. Chapter 4, by Morley, S. Financial appraisal: sensitivity and probability analysis, Estates Gazette, London.

Eatwell, J., Milgate, M. and Newman, P. (1990), *The New Palgrave: Utility and Probability*, Macmillan, Basingstoke.

Evans, A. H. (1992), Monte Carlo Analysis: a practical application to development appraisal. *Journal of Property Finance*, 3 (2), 271–281.

Fairley, W. and Jacoby, H. D. (1975), Investment analysis using the probability distribution of the internal rate of return. *Management Science*, 21 (12), 1428–1437.

Farrell, P. B. (1969), Computer aided financial risk simulation. *Appraisal Journal*, 37 (1), 58–73.

French, S. (1986), *Decision Theory: An Introduction to the Mathematics of Rationality*, Ellis Horwood, Chichester.

French, S. (ed.) (1989), *Readings in Decision Analysis*, Chapman and Hall, London.

Friedman, M. and Savage, L. J. (1948), The utility analysis of choices involving risk. *Journal of Political Economy*, 4, 279–304.

Galbraith, J. K. (1977), *The Age of Uncertainty*, BBC and Andre Deutsch Ltd, London.

Greer, G. E. (1979), *The Real Estate Investment Decision*, D. C. Heath, Farnborough, Chapters 10–14.

Hampton, J., Moore, P. G. and Thomas, H. (1973), Subjective probability and its measurement. *Journal of the Royal Statistical Society*, A136 (1), 21–42.

Hemmer, E. H. and Fisher, J. D. (1978), Dealing with uncertainty in income valuation. *Appraisal Journal*, 46 (2), 230–244.

Hertz, D. B. (1964), Risk analysis in capital investment. *Harvard Business Review*, 42 (1), 95–106.

Hertz, D. B. (1968), Investment policies that pay off. *Harvard Business Review*, 46 (1), 96–108.

Hertz, D. B. and Thomas, H. (1983), *Risk Analysis and its Applications*, J. Wiley, Chichester.

Hertz, D. B. and Thomas, H. (1984), *Practical Risk Analysis: An Approach through Case Histories*, J. Wiley, Chichester.

Hespos, R. F. and Strassman, P. A. (1965), Stochastic decision trees for the analysis of investment decisions. *Management Science*, B, 11 (10), 244–259.

Hillier, F. S. (1963), The derivation of probabilistic information for the evaluation of risky investments. *Management Science*, 9 (3), 443–457.

Hillier, F. S. (1965), Supplement to 'The derivation of probabilistic information for evaluation of risky investments'. *Management Science*, 11 (3), 485–487.

HMSO Report (1975), *Commercial Property Development*, First Report of the Government's Advisory Group on Commercial Property Development, HMSO, London.

HMSO Report (1980), *The Committee to Review the Functioning of Financial Institutions*, Cmnd 7937, HMSO, London.

Holloway, C. A. (1979), *Decision Making under Uncertainty: Models and Choices*, Prentice Hall, New York.

Huber, G. P. (1974), Methods for quantifying probabilities and multivariate utilities. *Decision Sciences*, 5 (3), 430–458.

Hull, J. C. (1977), Dealing with dependence in risk simulation. *Operational Research Quarterly*, 28 (1) ii, 201–213.

Hull, J. C. (1977), The input to and output from risk evaluation models. *European Journal of Operational Research*, 1 (6), 368–375.

Hull, J. C. (1977), Reducing the number of probabilistic variables in risk simulation. *Omega*, 5 (5), 605–608.

Hull, J. C. (1978), The interpretation of the output from a sensitivity analysis in investment appraisal. *Journal of Business Finance and Accounting*, 5 (1), 109–121.

Hull, J. C. (1978), The accuracy of the means and standard deviations of subjective probability distributions. *Journal of the Royal Statistical Society*, A141 (1), 79–85.

Hull, J. C. (1980), *The Evaluation of Risk in Business Investment*, Pergamon, Oxford.

Hull, J. C., Moore, P. G. and Thomas, H. (1973), Utility and its measurement. *Journal of the Royal Statistical Society*, A136 (2), 226–247.

Key, A, Zarkesh, F., MacGregor, B. and Nanthakumaran, N. (1994), *Understanding the Property Cycle*. Royal Institution of Chartered Surveyors, London.

Keeney, R. L. and Raiffa, H. (1976), *Decisions with Multiple Objectives: Preferences and Trade offs*. John Wiley, New York.

Kotz, S. and Stroup, D. A. (1983), *Educated Guessing: How to Cope in an Uncertain World*. Marcel Dekker, New York.

Kryzanowski, L., Lustig, P. and Schwab, B. (1973), Monte Carlo simulation and capital expenditure decisions: a case study. *Engineering Economist*, 18 (1), 31–48.

Lewellen, W. G. and Long, M. S. (1972), Simulation versus single value estimates in capital expenditure analysis. *Decision Analysis*, 3, 19–33.

Magee, J. F. (1964), Decision trees for decision making. *Harvard Business Review*, 42 (4), 126–138.

Magee, J. F. (1964), How to use decision trees in capital investment. *Harvard Business Review*, 42 (5), 79–86.

Marshall, P. and Kennedy, C. (1992), Development valuation techniques. *Journal of Property Valuation and Investment*, 11 (1), 57–66.

Martin, W. B. (1978), A risk analysis rate of return model for evaluating income producing real estate investments. *Appraisal Journal*, 46 (3), 424–442.

Mavrides, L. P. (1979), Decision analysis for real estate and equipment leasing. *Real Estate Appraiser and Analyst*, 45 (4), 39–48.

Mollart, R. G. (1988), Monte Carlo simulation using LOTUS 123. *Journal of Property Valuation*, 6 (4), 419–433.

Mollart, R.G. (1994), Using @RISK for risk analysis. *Journal of Property Valuation and Investment*, 12 (3), 89–96.

Moore, P. G. (1977), The manager's struggles with uncertainty. *Journal of the Royal Statistical Socieety*, A140 (2), 129–165.

Moore, P. G. (1986), *The Business of Risk*, Cambridge UP, Cambridge.

Moore, P. G. and Thomas, H. (1975), Measuring uncertainty. *Omega*, 3 (6), 657–672.

Moore, P. G. and Thomas, H. (1988), *The Anatomy of Decisions*, 2nd edn, Penguin, London.

Orman, G. A. E. (1991), Decision analysis techniques for property investment. *Journal of Property Finance*, 2 (3), 403–409.

Pellatt, P. G. K. (1972), The analysis of real estate investments under uncertainty. *Journal of Finance*, 27 (2), 459–470.

Penny, P. E. (1982), Modern investment theory and real estate analysis. *Appraisal Journal*, 50 (1), 79–99.

Phyrr, S. A. (1973), A computer simulation model to measure the risk in real estate investment. *American Real Estate and Urban Economic Association Journal (AREUEA J.)*, 1 (1), 48–78.

Plender, J. (1982), *That's the Way the Money Goes*, Andre Deutsch, London.

Raftery, I. (1994), *Risk Analysis in Project Management*. E. & F. N. Spon, London.

Raiffa, H. (1968), *Decision Analysis: Introductory Lectures on Choice under Uncertainty*. Addison Wesley, Reading, Mass.

Ratcliff, R. U. and Schwab, B. (1970), Contemporary decision theory and real estate investment. *Appraisal Journal*, 38 (2), 165–187.

Ratcliffe, J. (1973), Uncertainty and risk in development appraisal. *Estates Gazette*, 227, 603.

Robichek, A. A. (1975), Interpreting the results of risk analysis. *Journal of Finance*, 30 (5), 1384–1386.

Ross Goobey, A. (1992), *Bricks and Mortals*, Century Business, London.

Salazar, R. C. and Sen, S. K. (1968), A simulation model of capital budgeting under uncertainty. *Management Science*, 15B (4), 161–179.

Savage, L. J. (1954), *The Foundations of Statistics*, Wiley, New York.

Schlaifer, R. O. (1969), *Analysis of Decisions Under Uncertainty*, McGraw Hill, New York.

Schlaifer, R. O. (1971), *Computer Programs for Elementary Decision Analysis*, Division of Research, Harvard University.

Shackle, G. L. S. (1961), *Decision, Order and Time in Human Affairs*, Cambridge University Press, Cambridge.

Smith, J. Q. (1988), *Decision Analysis: A Bayesian Approach*. Chapman and Hall, London.

Spetzler, C. S. and Stael Von Holstein, C.-A. S. (1975), Probability encoding in decision analysis. *Management Science*, 22 (3), 340–358.

Stevenson, H. H. and Jackson, B. B. (1977), Large scale real estate investment-understanding the risks through modelling. *Appraisal Journal*, 45 (3), 366–382.

Swalm, R. O. (1966), Utility theory-insights into risk theory. *Harvard Business Review*, 44 (6), 123–136.

Tversky, A. (1974), Assessing uncertainty. *Journal of the Royal Statistical Society*, B36 (2), 148–159.

Von Neumann, J. and Morganstern, O. (1948), *The Theory of Games and Economic Behaviour*, 2nd edn, Princeton University Press.

Wagle, B. (1967), A statistical analysis of risk in capital investment projects. *Operational Research Quarterly*, 18 (1), 13–33.

Winkler, R. L. (1967), The assessment of prior distributions in Bayesian analysis. *Journal of the American Statistical Society*, 62, 776–800.

Wofford, L. E. (1978), A simulation approach to the appraisal of income producing real estate. *American Real Estate and Urban Economic Association Journal (AREUEA J.)*, 6 (4), 370–394.

Wofford, L. E. (1979), Incorporating uncertainty into the data program. *Real Estate Appraiser and Analyst*, 45 (3), 30–38.

Wood, E. (1977), *Property and Building Appraisal in Uncertainty*. Occasional Paper, Department of Surveying, Liverpool Polytechnic, Liverpool.

Woods, D. H. (1966), Improving estimates that involve uncertainty. *Harvard Business Review*, 44, 91–98.

Index

Printed in the United Kingdom
by Lightning Source UK Ltd.
126999UK00001B/256-318/A